GETTING STARTED IN ELECTRONICS

BY FORREST M. MIMS III

COPYRIGHT © 1983, 2000 BY FORREST M. MIMS, III
ALL RIGHTS RESERVED

Published for Forrest M. Mims III by:
MASTER PUBLISHING, INC.
7101 N.. Ridgeway Ave.
Lincolnwood, IL 60712
847-763-0916 (voice)
847-763-0918 (fax)
masterpubl@aol.com (e-mail)

Visit Master Publishing
on the Internet at:
www.masterpublishing.com

REGARDING THESE BOOK MATERIALS
Reproduction, publication, or duplication of this book, or any part thereof, in any manner, mechanically, electronically, or photographically, is prohibited without the express written permission of the Author and Publisher. For permission and other rights under this copyright, write Master Publishing, Inc.

The Author, Publisher, and Seller assume no liability with respect to the use of the information contained herein.

First edition:	Second edition:	Third edition:	Third edition:
Fifteen printings	First printing–1998	Second printing–2003	Third printing–2006

THIS BOOK IS FOR THE ENTERTAINMENT AND EDIFICATION OF ITS READERS. WHILE REASONABLE CARE HAS BEEN EXERCISED WITH RESPECT TO ITS ACCURACY, THE AUTHOR AND RADIO SHACK ASSUME NO RESPONSIBILITY FOR ERRORS, OMISSIONS OR SUITABILITY OF ITS CONTENTS FOR ANY APPLICATION. NEITHER DO WE ASSUME ANY LIABILITY FOR ANY DAMAGES RESULTING FROM USE OF INFORMATION IN THIS BOOK. IT IS YOUR RESPONSIBILITY TO DETERMINE IF USE, MANUFACTURE OR SALE OF ANY DEVICE THAT INCORPORATES INFORMATION IN THIS BOOK INFRINGES ANY PATENTS, COPYRIGHTS OR OTHER RIGHTS.

CAUTION: THIS BOOK INCLUDES SEVERAL REFERENCES TO ELECTRICAL SAFETY WHICH MUST BE HEEDED. IT IS ESSENTIAL THAT CAREFUL SUPERVISION BE GIVEN CHILDREN WORKING WITH LINE-POWERED ELECTRONIC CIRCUITS AND SOLDERING IRONS.

DUE TO THE MANY CUSTOMER INQUIRIES RECEIVED BY RADIOSHACK AND THE AUTHOR, IT IS IMPOSSIBLE TO ANSWER REQUESTS FOR ADDITIONAL INFORMATION (CUSTOM CIRCUIT DESIGNS, TECHNICAL ADVICE, TROUBLESHOOTING ASSISTANCE, ETC.). BUT THOUGH WE CANNOT ACKNOWLEDGE INDIVIDUAL INQUIRIES, WE WILL BE HAPPY TO RECEIVE ANY COMMENTS, IMPRESSIONS OR SUGGESTIONS.

THANKS IN ADVANCE TO THOSE OF YOU WHO WRITE! BUT PLEASE REMEMBER WE ARE UNABLE TO GIVE YOU A PERSONAL REPLY.

CONTENTS

GETTING STARTED IN ELECTRONICS	5
1. ELECTRICITY	6
PUTTING ELECTRICITY TO WORK	7
BACK TO BASICS	8-9
STATIC ELECTRICITY	10-12
ELECTRICAL CURRENT	13
DIRECT CURRENT ELECTRICITY	14-17
ALTERNATING CURRENT ELECTRICITY	18
MEASURING AC AND DC	19
ELECTRICAL CIRCUITS	20-21
PULSES, WAVES, SIGNALS AND NOISE	22-23
2. ELECTRONIC COMPONENTS	24
WIRE AND CABLE	24
SWITCHES	25
RELAYS	26
MOVING COIL METERS	26
MICROPHONES AND SPEAKERS	27
RESISTORS	28-31
CAPACITORS	32-35
RESISTOR AND CAPACITOR APPLICATIONS	36-37
COILS	38-39
TRANSFORMERS	40-41
3. SEMICONDUCTORS	42
SILICON	42-43
THE DIODE	44-47
THE TRANSISTOR	48
BIPOLAR TRANSISTORS	48-51
FIELD-EFFECT TRANSISTORS	52-56
THE UNIJUNCTION TRANSISTOR	57
THE THYRISTOR	58
SILICON-CONTROLLED RECTIFIERS	58-59
TRIACS	60-61
4. PHOTONIC SEMICONDUCTORS	62
LIGHT	62-63
OPTICAL COMPONENTS	64
HOW CONVEX LENSES ARE USED	65
SEMICONDUCTOR LIGHT SOURCES	66
LIGHT EMITTING DIODES	66-69
SEMICONDUCTOR LIGHT DETECTORS	70
PHOTORESISTIVE LIGHT DETECTORS	70-71
PN JUNCTION LIGHT DETECTORS	72
PHOTODIODES	72-73
PHOTOTRANSISTORS	74-75
PHOTOTHYRISTORS	76
SOLAR CELLS	77

5. INTEGRATED CIRCUITS — 78-79
6. DIGITAL INTEGRATED CIRCUITS — 80
- MECHANICAL SWITCH GATES — 80
- THE BINARY CONNECTION — 81
- DIODE GATES — 82
- TRANSISTOR GATES — 83
- GATE SYMBOLS — 84-85
- DATA HIGHWAYS — 85
- HOW GATES ARE USED — 86
- COMBINATIONAL LOGIC CIRCUITS — 86-87
- SEQUENTIAL LOGIC CIRCUITS — 88-89
- A COMBINATIONAL-SEQUENTIAL LOGIC SYSTEM — 90
- DIGITAL IC FAMILIES — 91

7. LINEAR INTEGRATED CIRCUITS — 92
- THE BASIC LINEAR CIRCUIT — 92
- OPERATIONAL AMPLIFIERS — 93
- TIMERS — 94
- FUNCTION GENERATORS — 94
- VOLTAGE REGULATORS — 95
- OTHER LINEAR IC's — 95

8. CIRCUIT ASSEMBLY TIPS — 96
- TEMPORARY CIRCUITS — 96
- PERMANENT CIRCUITS — 96-97
- HOW TO SOLDER — 98
- POWERING ELECTRONIC CIRCUITS — 99

9. 100 ELECTRONIC CIRCUITS — 100
- DIODE CIRCUITS — 101
  - SMALL SIGNAL DIODES AND RECTIFIERS — 101-102
  - ZENER DIODE CIRCUITS — 103
- TRANSISTOR CIRCUITS — 104
  - BIPOLAR TRANSISTOR CIRCUITS — 104-105
  - JUNCTION FET AND POWER MOSFET CIRCUITS — 106-107
- UNIJUNCTION TRANSISTOR CIRCUITS — 108-109
- THYRISTOR CIRCUITS — 110
  - SCR CIRCUITS — 110
  - TRIAC CIRCUITS — 111
- PHOTONIC CIRCUITS — 112
  - LIGHT EMITTING DIODE (LED) CIRCUITS — 112-113
  - SEMICONDUCTOR LIGHT DETECTOR CIRCUITS — 114-115
- DIGITAL IC CIRCUITS — 116
  - TTL CIRCUITS — 116-117
  - CMOS CIRCUITS — 118-121
- LINEAR IC CIRCUITS — 122
  - OPERATIONAL AMPLIFIER (OP-AMP) CIRCUITS — 122-123
  - COMPARATOR CIRCUITS — 124
  - VOLTAGE REGULATOR CIRCUITS — 125
  - TIMER CIRCUITS — 126-127

INDEX — 128

GETTING STARTED IN ELECTRONICS

Welcome to the world of electronics, one of the fastest growing of today's "high-tech" fields and an educational and entertaining hobby. This book will take you from static electricity to solid-state electronics. Along the way we'll cover electricity, electronic components and integrated circuits (IC's). Chapters 3-7 show how components are used to form electronic circuits. Chapter 9 gives plans for 100 circuits, each of which I've built and tested. "Page arrows" (▭▷) throughout the book refer you to related topics in future chapters (like working versions of many example circuits in chapters 3-7). I hope you find this book useful, educational and, especially, fun! *Forrest M. Mims III*

GOING FURTHER IN ELECTRONICS

I hope this book encourages you to go further in electronics. You can begin by building the hundreds of circuits in the "Engineer's Mini-Notebook" series I have written for RadioShack. You should also read hobby electronics magazines. Remember this: You will learn more from building, testing and using electronic circuits than by simply reading about them. So be sure to actually build as many of the circuits in this book as you can. Questions? This book will raise lots of questions if you're serious about electronics. You can find answers to many of your questions in the "Engineer's Mini-Notebook" series and other RadioShack books. Be sure to see the books about electronics at the local library. Finally, be sure to check out the various hobby electronics news groups and resources — including RadioShack's — on the internet and world wide web.

A SPECIAL NOTE TO EDUCATORS

Since this book was first published in 1983, it has helped many students develop award winning science projects. Some teachers assign various sections of the book. Others others use the entire book as the text for courses on basic electronics. Thanks to RadioShack's solderless, modular sockets, you and your students should be able to assemble test versions of virtually every circuit in chapter 9 ("100 Electronic Circuits"). Incidentally, volume buyers should inquire at RadioShack about volume discounts on this book and RadioShack parts. Also see RadioShack's catalog. (Price discounts may not apply at RadioShack dealers and franchise stores.)

1. ELECTRICITY

The only difference between a bolt of lightning and the spark between your finger and a doorknob on a dry day is quantity. BOTH ARE ELECTRICITY. Benjamin Franklin first confirmed this with his famous kite experiment.

To those who fly a kite in the rain, you better say "bye!" 'cause it's NOT very sane.

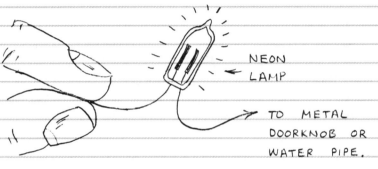

Here's a neat way to "see" electricity without being zapped: grasp one lead from a neon lamp, walk across a carpet while wearing hard soled shoes and touch the second lead from the lamp to a metal object. The lamp will flash (unless the relative humidity is high).

Of course, you cannot "see" electricity! You see its EFFECT upon air and the neon in the lamp. The EFFECTS of electricity which can be seen are MANY. Here are some more:

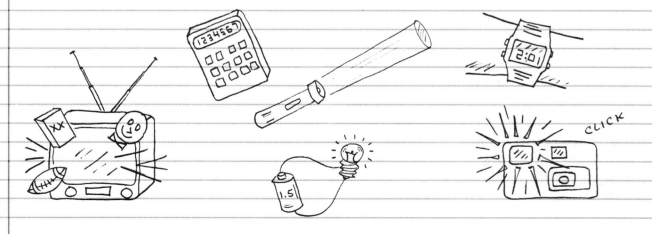

PUTTING ELECTRICITY TO WORK

All matter has electrical properties. That's why scientists over the past few centuries have been able to invent HUNDREDS of gadgets that generate, store, control and switch electricity. These devices have combined to carry us into...

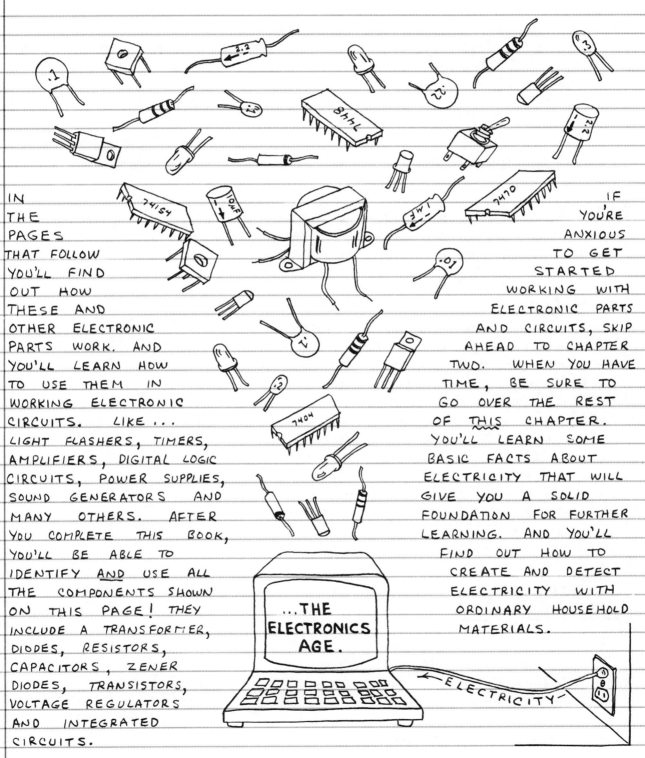

In the pages that follow you'll find out how these and other electronic parts work. And you'll learn how to use them in working electronic circuits. Like... light flashers, timers, amplifiers, digital logic circuits, power supplies, sound generators and many others. After you complete this book, you'll be able to identify AND use all the components shown on this page! They include a transformer, diodes, resistors, capacitors, zener diodes, transistors, voltage regulators and integrated circuits.

If you're anxious to get started working with electronic parts and circuits, skip ahead to Chapter Two. When you have time, be sure to go over the rest of this chapter. You'll learn some basic facts about electricity that will give you a solid foundation for further learning. And you'll find out how to create and detect electricity with ordinary household materials.

7

BACK TO BASICS

ELECTRICITY IS AN ESSENTIAL INGREDIENT OF MATTER. THE BEST WAY TO UNDERSTAND THE NATURE OF ELECTRICITY IS TO EXAMINE THE SMALLEST COMPONENT OF EVERY ELEMENT, THE ATOM.

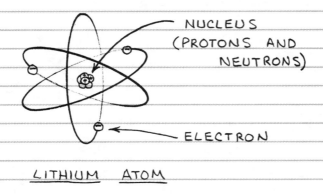

THIS IS A LITHIUM ATOM. THE THIRD SIMPLEST ATOM AFTER HYDROGEN AND HELIUM, LITHIUM ATOMS HAVE 3 <u>ELECTRONS</u> THAT ENCIRCLE A NUCLEUS OF 3 <u>PROTONS</u> AND 4 <u>NEUTRONS</u>.

- ⊖ ELECTRONS HAVE A <u>NEGATIVE</u> ELECTRICAL CHARGE.
- ⊕ PROTONS HAVE A <u>POSITIVE</u> ELECTRICAL CHARGE.
- ◯ NEUTRONS HAVE <u>NO</u> ELECTRICAL CHARGE.

☐ **IONS** — NORMALLY AN ATOM HAS AN EQUAL NUMBER OF ELECTRONS AND PROTONS. THE CHARGES CANCEL TO GIVE THE ATOM <u>NO</u> NET ELECTRICAL CHARGE. IT'S POSSIBLE TO DISLODGE ONE OR MORE ELECTRONS FROM MOST ATOMS. THIS CAUSES THE ATOM TO HAVE A NET POSITIVE CHARGE. IT'S THEN CALLED A <u>POSITIVE ION</u>. IF A STRAY ELECTRON COMBINES WITH A NORMAL ATOM, THE ATOM HAS A NET NEGATIVE CHARGE AND IS CALLED A <u>NEGATIVE ION</u>.

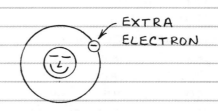

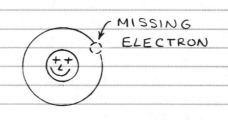

☐ **ELECTRONS** — <u>FREE</u> ELECTRONS CAN MOVE AT HIGH SPEED THROUGH METALS, GASES AND A VACUUM. OR THEY CAN REST ON A SURFACE.

☐ **MORE ABOUT FREE ELECTRONS** — MANY TRILLIONS OF ELECTRONS CAN REST ON A SURFACE OR TRAVEL THROUGH SPACE OR MATTER AT NEAR THE SPEED OF LIGHT (186,000 MILES PER SECOND)!

RESTING ELECTRONS   MOVING ELECTRONS

☐ **RESTING ELECTRONS** — A GROUP OF NEGATIVE ELECTRONS ON A SURFACE CAUSES THE SURFACE TO BE NEGATIVELY CHARGED. SINCE THE ELECTRONS ARE NOT MOVING, THE SURFACE CAN BE SAID TO HAVE A <u>NEGATIVE STATIC ELECTRICAL CHARGE</u>.

☐ **MOVING ELECTRONS** — A STREAM OF MOVING ELECTRONS IS CALLED AN <u>ELECTRICAL CURRENT</u>. RESTING ELECTRONS CAN QUICKLY FORM AN ELECTRICAL CURRENT IF PLACED NEAR A CLUSTER OF POSITIVE IONS. THE POSITIVELY CHARGED IONS WILL ATTRACT THE ELECTRONS WHICH WILL RUSH IN TO FILL THE "HOLES" OR VOIDS LEFT BY THE MISSING ELECTRONS.

MISSING ELECTRON ("HOLE")
ELECTRON ORBIT
ELECTRON

☐ **MISSING ELECTRONS** — MECHANICAL FRICTION, LIGHT, HEAT OR A CHEMICAL REACTION MAY REMOVE ELECTRONS FROM A SURFACE. THIS CAUSES THE SURFACE TO BE POSITIVELY CHARGED. SINCE THE POSITIVELY CHARGED ATOMS ARE AT REST, THE SURFACE CAN BE SAID TO HAVE A <u>POSITIVE STATIC ELECTRICAL CHARGE</u>.

FRICTION, LIGHT, HEAT, CHEMICALS

POSITIVE IONS WITH POSITIVE STATIC ELECTRICAL CHARGE.

STATIC ELECTRICITY

You generate static electricity every time you walk across a carpet, pull tape from a roll, remove your clothing or dry clothes in a drier. Much of the time you don't even realize it unless the air is dry and the static charge suddenly crackles, pops and flashes its way to a new home. These static charges are caused by <u>MECHANICAL FRICTION</u>. Back in 600 B.C., Thales of Greece experimented with the static electricity produced when amber is rubbed with wool.

☐ **AMBER** — Once upon a time sap flowing from trees hardened into clear golden nodules which were eventually buried in the earth. Sometimes, before it hardened into amber, the sticky sap entombed bits of plant matter, insects and even droplets of water! A kind of natural casting plastic, amber is easily electrified by friction. It then attracts bits of paper.

<u>FAMOUS FACT</u>: The electron is named after the Greek word for amber!

☐ **ELECTRIFIED PLASTIC AND GLASS** — Run a plastic comb through your hair on a dry day and you'll transfer electrons from your hair to the comb. Rub a glass rod with silk or the synthetic fibers of a paint brush and you'll remove electrons from the glass. Both the negatively charged comb and the positively charged glass rod will, like amber, attract bits of paper. You can electrify or charge many materials by rubbing them with fur, wool, etc. Metal? No, the charge leaks away.

<u>COMB</u> (after stroking hair) — negative charge

<u>GLASS ROD</u> (rubbed with silk) — positive charge

bits of paper

10

☐ **OPPOSITE AND LIKE CHARGES** — HOW DO WE KNOW THE COMB AND GLASS ROD HAVE OPPOSITE CHARGES? A FUNDAMENTAL RULE OF ELECTRICITY IS <u>LIKE CHARGES REPEL AND UNLIKE CHARGES ATTRACT</u>. HERE'S AN EXPERIMENT THAT PROVES THE RULE AND ANSWERS THE QUESTION:

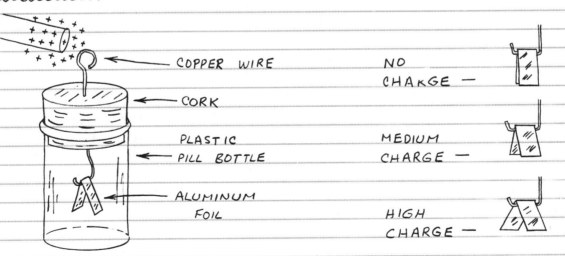

* USE FOAM PLASTIC PACKING MATERIAL.

<u>REMEMBER</u>: UNLIKE CHARGES ATTRACT —

LIKE CHARGES REPEL —

☐ **THE ELECTROSCOPE** — THE FIRST INSTRUMENT DESIGNED TO DETECT AND MEASURE STATIC ELECTRICITY WAS THE <u>ELECTROSCOPE</u>. YOU CAN EASILY MAKE ONE.

- COPPER WIRE
- CORK
- PLASTIC PILL BOTTLE
- ALUMINUM FOIL

NO CHARGE —

MEDIUM CHARGE —

HIGH CHARGE —

BE SURE THE FOLDED FOIL STRIP IS CLEAN AND DRY. WHEN YOU TOUCH A CHARGED OBJECT TO THE WIRE, THE TWO HALVES OF THE FOIL STRIP WILL BE GIVEN THE SAME CHARGE AND WILL THEREFORE FLY APART.

☐ CONDUCTORS AND INSULATORS— YOU CAN USE YOUR ELECTROSCOPE TO PROVE THAT ELECTRONS TRAVEL THROUGH SOME MATERIALS BUT NOT OTHERS. <u>HINT</u>: TRY THIS ON A <u>DRY</u> DAY! ELECTRONS CAN TRAVEL THROUGH MOIST AIR SO THE CHARGE ON YOUR ELECTROSCOPE WILL QUICKLY LEAK AWAY ON HUMID DAYS.

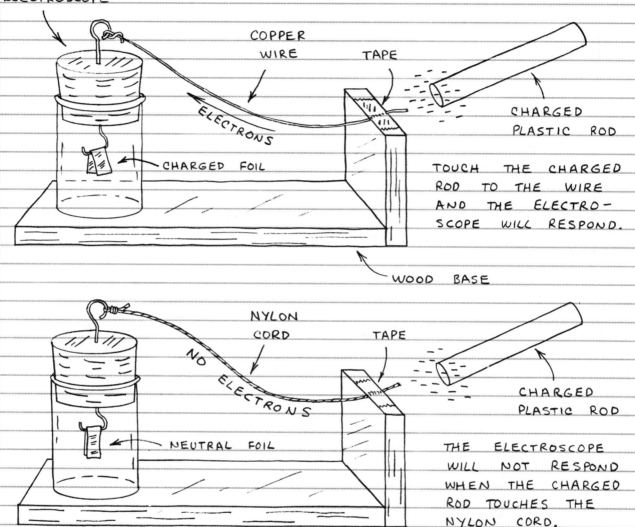

THIS DEMONSTRATION SHOWS THAT ELECTRONS CAN TRAVEL THROUGH SOME MATERIALS BUT NOT OTHERS. MATERIALS THROUGH WHICH ELECTRONS TRAVEL ARE <u>CONDUCTORS</u>. MATERIALS THROUGH WHICH ELECTRONS TRAVEL POORLY OR NOT AT ALL ARE <u>INSULATORS</u>.

<u>CONDUCTORS</u> INCLUDE SILVER, GOLD, IRON, COPPER, ETC.

<u>INSULATORS</u> INCLUDE GLASS, PLASTIC, RUBBER, WOOD, ETC.

ELECTRICAL CURRENT

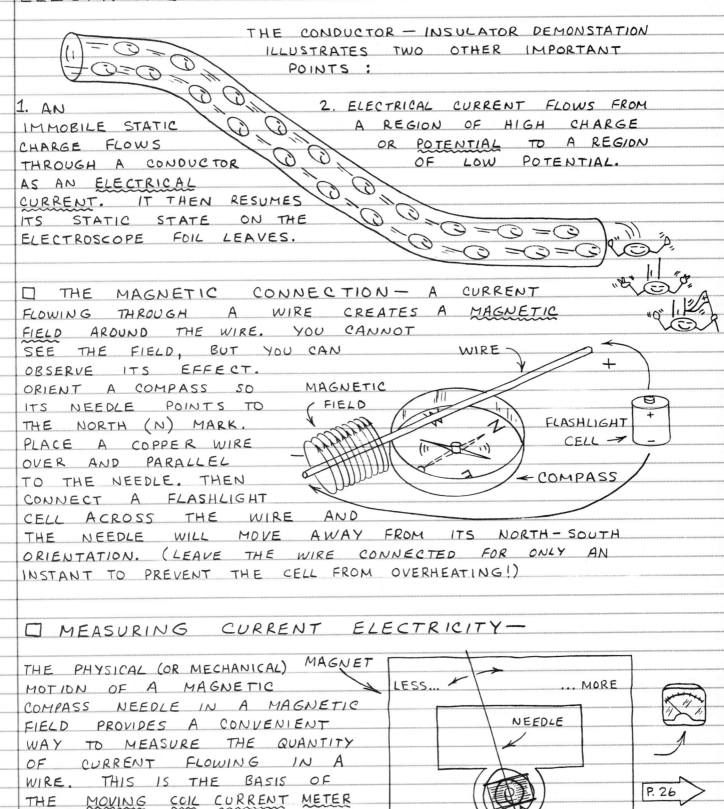

THE CONDUCTOR — INSULATOR DEMONSTRATION ILLUSTRATES TWO OTHER IMPORTANT POINTS:

1. AN IMMOBILE STATIC CHARGE FLOWS THROUGH A CONDUCTOR AS AN <u>ELECTRICAL CURRENT</u>. IT THEN RESUMES ITS STATIC STATE ON THE ELECTROSCOPE FOIL LEAVES.

2. ELECTRICAL CURRENT FLOWS FROM A REGION OF HIGH CHARGE OR <u>POTENTIAL</u> TO A REGION OF LOW POTENTIAL.

☐ THE MAGNETIC CONNECTION— A CURRENT FLOWING THROUGH A WIRE CREATES A <u>MAGNETIC FIELD</u> AROUND THE WIRE. YOU CANNOT SEE THE FIELD, BUT YOU CAN OBSERVE ITS EFFECT. ORIENT A COMPASS SO ITS NEEDLE POINTS TO THE NORTH (N) MARK. PLACE A COPPER WIRE OVER AND PARALLEL TO THE NEEDLE. THEN CONNECT A FLASHLIGHT CELL ACROSS THE WIRE AND THE NEEDLE WILL MOVE AWAY FROM ITS NORTH-SOUTH ORIENTATION. (LEAVE THE WIRE CONNECTED FOR ONLY AN INSTANT TO PREVENT THE CELL FROM OVERHEATING!)

☐ MEASURING CURRENT ELECTRICITY—

THE PHYSICAL (OR MECHANICAL) MOTION OF A MAGNETIC COMPASS NEEDLE IN A MAGNETIC FIELD PROVIDES A CONVENIENT WAY TO MEASURE THE QUANTITY OF CURRENT FLOWING IN A WIRE. THIS IS THE BASIS OF THE <u>MOVING COIL</u> CURRENT METER USED IN THE ANALOG MULTIMETER. TO PROVIDE HIGH SENSITIVITY, THE WIRE IS WRAPPED AS A COIL.

DIRECT CURRENT ELECTRICITY

An electrical current can flow in either of two directions through a conductor. If it flows in only one direction, whether steadily or in pulses, it's called <u>direct current</u> (DC). It's important to be able to specify the quantity and power of a direct current. Here are the key terms:

☐ **CURRENT (I)** — Current is the quantity of electrons passing a given point. The unit of current is the <u>ampere</u>. One ampere is 6,280,000,000,000,000,000 (6.28×10^{18}) electrons passing a point in one second.

☐ **VOLTAGE (V or E)** — Voltage is electrical pressure or force. Voltage is sometimes referred to as <u>potential</u>. Voltage drop is the difference in voltage between the two ends of a conductor through which current is flowing. If we compare current to water flowing through a pipe, then voltage is the water pressure.

☐ **POWER (P)** — The work performed by an electrical current is called power. The unit of power is the <u>watt</u>. The power of a direct current is its voltage times its current.

☐ **RESISTANCE (R)** — Conductors are not perfect. They resist to some degree the flow of current. The unit of resistance is the <u>ohm</u> (Ω). A potential difference of one volt will force a current of one ampere through a resistance of one ohm. The resistance of a conductor is its voltage drop divided by the current flowing through the conductor.

☐ **MR. OHM'S LAW** — Given any two of the above, you can find the other two using these formulas known as <u>Ohm's Law</u>:

$$V = I \times R$$
$$I = V / R$$
$$R = V / I$$
$$P = V \times I \text{ (or) } I^2 \times R$$

We'll refer to Ohm's Law later in this book...

☐ **SUMMING UP** — This is the "water analogy":

- Water level (voltage)
- Tap (resistance)
- Stream (current)
- Rotating turbine (power)

USING DIRECT CURRENT

THERE ARE SO MANY USES FOR DIRECT CURRENT ELECTRICITY NO SINGLE BOOK CAN DESCRIBE THEM ALL. HERE'S A PAGE OF SEVERAL DESIGNED AROUND A SINGLE WIRE COIL YOU CAN EASILY MAKE FROM A 1-½ TO 3-INCH SECTION OF A SODA STRAW AND AT LEAST 30-FEET OF 30 GAUGE, LACQUER COATED WIRE. SECURE THE COIL IN PLACE WITH TAPE. REMOVE INSULATION FROM ENDS OF COIL WITH FINE SAND PAPER.

☐ ELECTROMAGNET — INSERT A STEEL NAIL IN THE COIL, CONNECT THE LEADS TO A 9-VOLT BATTERY, AND THE NAIL WILL BECOME A MAGNET UNTIL THE POWER IS DISCONNECTED. (IT MAY RETAIN SOME MAGNETISM.)

☐ SOLENOID — THIS IS A "SUCKING MAGNET." APPLY POWER TO COIL AND NAIL WILL BE PULLED RAPIDLY INSIDE.

☐ MOTOR — MAYBE NOT YOUR IDEA OF A MOTOR, BUT THIS ELEGANT APPARATUS QUALIFIES UNDER THE DICTIONARY DEFINITION. USE A LIGHT WEIGHT NAIL. SECURE ONE COIL LEAD TO NAIL. ADJUST HEIGHT OF COIL UNTIL NAIL JUMPS UP AND DOWN.

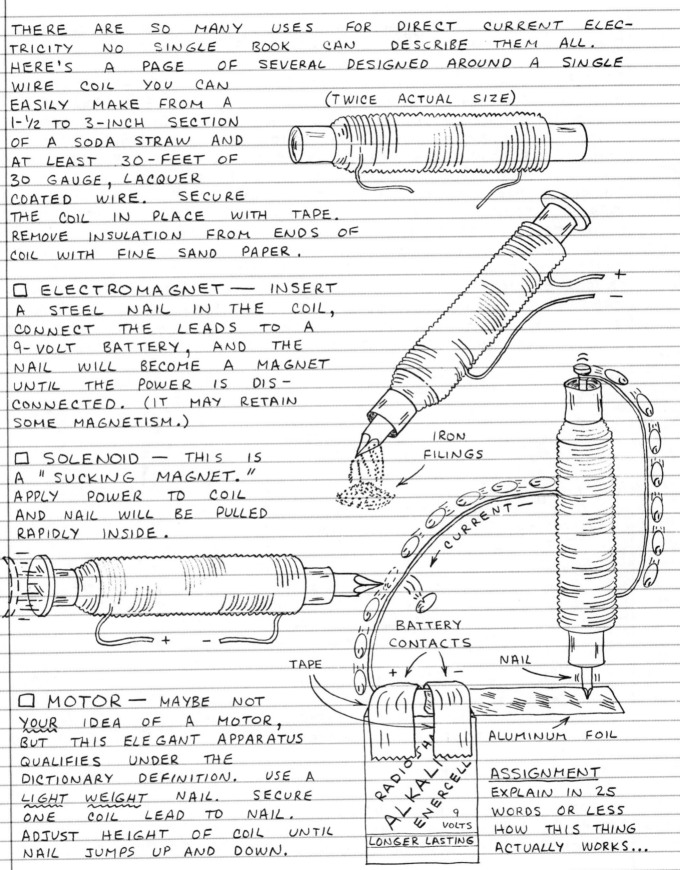

(TWICE ACTUAL SIZE)

IRON FILINGS

CURRENT

BATTERY CONTACTS

TAPE

NAIL

ALUMINUM FOIL

RADIOSH ALKALI ENERCELL 9 VOLTS LONGER LASTING

ASSIGNMENT
EXPLAIN IN 25 WORDS OR LESS HOW THIS THING ACTUALLY WORKS...

15

MAKING DIRECT CURRENT ELECTRICITY

A SURPRISING NUMBER OF WAYS EXIST FOR PRODUCING DIRECT CURRENT. HERE ARE THE BIGGIES:

☐ **CHEMICAL GENERATORS** — ELECTROLYTES ARE CHEMICAL SOLUTIONS THAT CONTAIN MANY IONS. FOR EXAMPLE, DISSOLVE TABLE SALT IN WATER AND THE SALT WILL BREAK DOWN INTO POSITIVE SODIUM IONS AND NEGATIVE CHLORINE IONS. IF TWO DISSIMILAR METAL PLATES ARE IMMERSED IN THE SALT SOLUTION, THE POSITIVE IONS WILL MIGRATE TOWARD ONE PLATE AND THE NEGATIVE IONS WILL MIGRATE TOWARD THE OTHER. IF THE TWO PLATES ARE CONNECTED TOGETHER BY A CONDUCTOR, A CURRENT WILL FLOW THROUGH THE SOLUTION (AS IONS) AND THE CONDUCTOR (AS ELECTRONS). THIS KIND OF GENERATOR IS CALLED A WET CELL. CELLS IN WHICH THE ELECTROLYTE IS ABSORBED BY PAPER OR FORMED INTO A PASTE ARE CALLED DRY CELLS. HERE ARE SOME CHEMICAL GENERATORS YOU CAN MAKE. HAVE FUN!

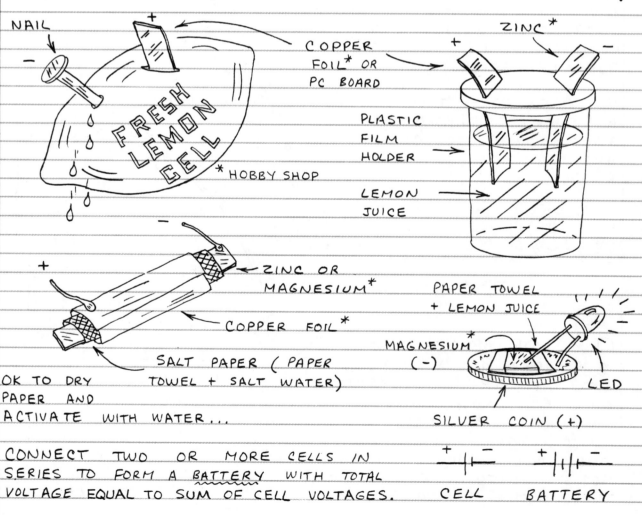

CONNECT TWO OR MORE CELLS IN SERIES TO FORM A BATTERY WITH TOTAL VOLTAGE EQUAL TO SUM OF CELL VOLTAGES.

16

☐ ELECTROMAGNETIC GENERATORS— A CURRENT FLOWING THROUGH A CONDUCTOR ESTABLISHES A MAGNETIC FIELD AROUND THE CONDUCTOR. THIS EFFECT WORKS BOTH WAYS SO THAT A CURRENT WILL FLOW IN A CONDUCTOR WHICH IS MOVED THROUGH A MAGNETIC FIELD. YOU CAN EASILY DEMONSTRATE ELECTROMAGNETIC CURRENT GENERATION WITH A COIL OF WIRE AND A SMALL MAGNET. (THE COIL SHOWN ON PAGE 15 WORKS FINE.) CONNECT THE LEADS OF THE COIL TO A METER DESIGNED TO SENSE MICROAMPERES. INSERT A STEEL NAIL THROUGH THE COIL AND STROKE THE MAGNET BACK AND FORTH ACROSS THE COIL. THE METER WILL INDICATE A FEW MICROAMPERES EACH STROKE. THE POLARITY (DIRECTION) OF THE CURRENT WILL REVERSE ON THE BACK STROKES. WANT A READY-MADE GENERATOR? JUST ROTATE THE SHAFT OF A SMALL DC MOTOR. MOST SUCH MOTORS WILL PRODUCE A POTENTIAL DIFFERENCE OF UP TO SEVERAL VOLTS! YOU CAN ADD A PROPELLER TO MAKE A WIND POWERED GENERATOR.

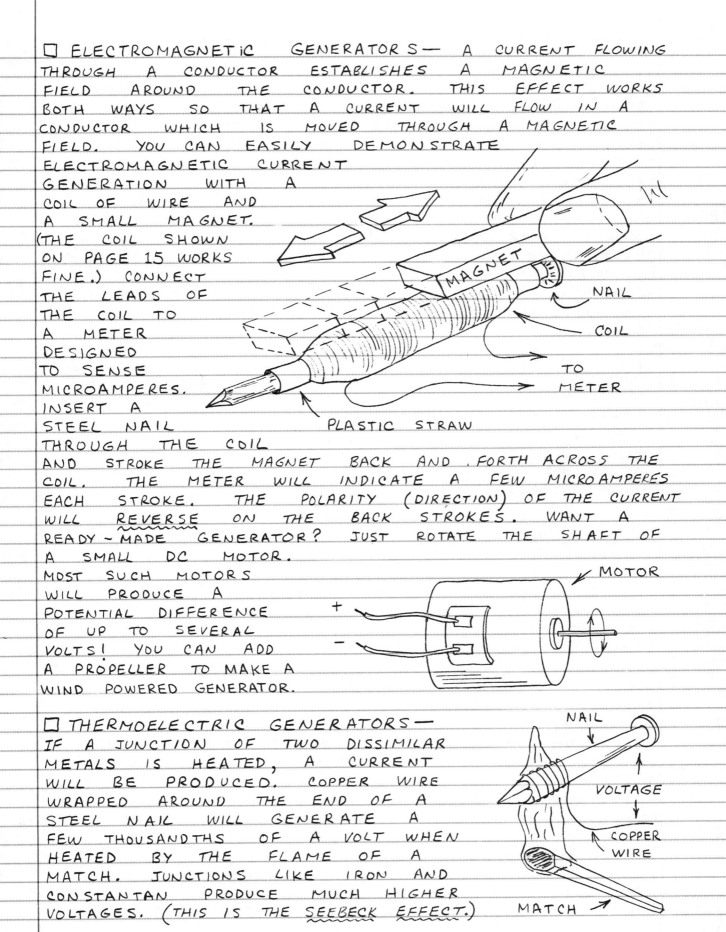

☐ THERMOELECTRIC GENERATORS— IF A JUNCTION OF TWO DISSIMILAR METALS IS HEATED, A CURRENT WILL BE PRODUCED. COPPER WIRE WRAPPED AROUND THE END OF A STEEL NAIL WILL GENERATE A FEW THOUSANDTHS OF A VOLT WHEN HEATED BY THE FLAME OF A MATCH. JUNCTIONS LIKE IRON AND CONSTANTAN PRODUCE MUCH HIGHER VOLTAGES. (THIS IS THE SEEBECK EFFECT.)

ALTERNATING CURRENT ELECTRICITY

Look back at the homemade coil and magnet "generator" on the preceeding page. When the magnet is stroked in one direction along the coil, electrons in the wire are moved in one direction and a DIRECT CURRENT is produced. On the back stroke, unless the magnet is moved away from the coil, the direction of current flow is reversed. Therefore, if the magnet is stroked back and forth along the coil, a current which alternates in direction or polarity is produced. It's called an ALTERNATING CURRENT. Alternating current (AC) is usually produced by rotating a coil in a magnetic field.

ROTATING COIL   VOLTAGE OUTPUT   AC SINE WAVE

☐ **SINE WAVE MEASUREMENT** — AC voltage is usually specified at a value equal to the DC voltage capable of doing the same work. For a sine wave this value is 0.707 times the peak voltage. It's called the RMS (root-mean-square) voltage. The peak voltage (or current) is 1.41 times the RMS value. Household line voltage is specified according to its RMS value. Therefore, a household voltage of 120-volts corresponds to a peak voltage of 120 × 1.41 or 169.2-volts.

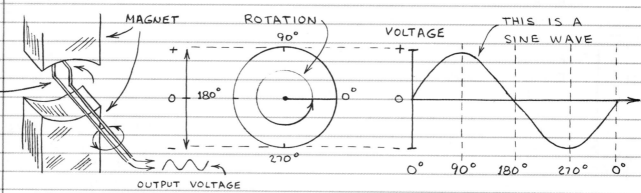

☐ **WHY AC IS USED** — AC is better suited than DC for transmission through long distance power lines. A wire carrying AC will induce a current in a nearby wire. This is the principle behind the TRANSFORMER.

MEASURING AC AND DC

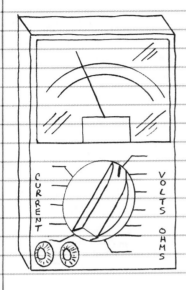

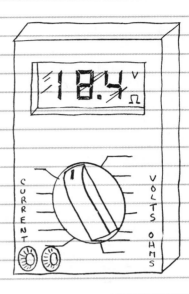

YOU CAN EASILY MEASURE AC AND DC VOLTAGE AND CURRENT WITH AN INSTRUMENT CALLED THE MULTIMETER. ANALOG MULTIMETERS USE A MOVING COIL METER. DIGITAL MULTIMETERS HAVE A DIGITAL READOUT. THE MULTIMETER IS THE SINGLE MOST IMPORTANT ELECTRONIC TEST INSTRUMENT.

☐ ANALOG MULTIMETER — LESS EXPENSIVE, SOMEWHAT LESS PRECISE THAN DIGITAL TYPES. BEST BY FAR FOR OBSERVING THE TREND OF A SLOWLY CHANGING VOLTAGE, CURRENT OR RESISTANCE.

☐ DIGITAL MULTIMETER — HIGHLY ACCURATE AND EASIER TO READ THAN ANALOG TYPES. BEST FOR FINDING THE PRECISE VALUE OF A VOLTAGE, CURRENT OR RESISTANCE.

☐ SUMMING UP MULTIMETERS — THEY'RE INDISPENSABLE! EVEN IF YOU HAVE ONLY A PASSING INTEREST YOU SHOULD CONSIDER BUYING ONE BECAUSE IT HAS MANY USES IN THE HOME, ON THE JOB AND WHEN WORKING WITH APPLIANCES AND MOTOR VEHICLES. IF YOU'RE SERIOUS ABOUT ELECTRONICS, CONSIDER BUYING A QUALITY HIGH-IMPEDANCE MULTIMETER THAT WILL HAVE LITTLE OR NO EFFECT ON THE DEVICE OR CIRCUIT YOU'RE MEASURING. IDEALLY, YOU SHOULD HAVE BOTH THE ANALOG AND DIGITAL TYPES.

ELECTRICAL SAFETY

ELECTRICITY CAN KILL! IF YOU WANT TO BE AROUND LONG ENOUGH TO ENJOY EXPERIMENTING WITH ELECTRONICS, ALWAYS TREAT ELECTRICITY WITH THE RESPECT IT DESERVES. WE'LL LOOK AT SAFETY AGAIN LATER.

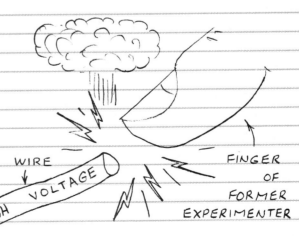

ELECTRICAL CIRCUITS

AN ELECTRICAL CIRCUIT IS ANY ARRANGEMENT THAT PERMITS AN ELECTRICAL CURRENT TO FLOW. A CIRCUIT CAN BE AS SIMPLE AS A BATTERY CONNECTED TO A LAMP OR AS COMPLICATED AS A DIGITAL COMPUTER.

☐ A BASIC CIRCUIT — THIS BASIC CIRCUIT CONSISTS OF A SOURCE OF ELECTRICAL CURRENT (A BATTERY), A LAMP AND TWO CONNECTION WIRES. THE PART OF A CIRCUIT WHICH PERFORMS WORK IS CALLED THE LOAD. HERE THE LOAD IS THE LAMP. IN OTHER CIRCUITS THE LOAD CAN BE A MOTOR, A HEATING ELEMENT, AN ELECTROMAGNET, ETC.

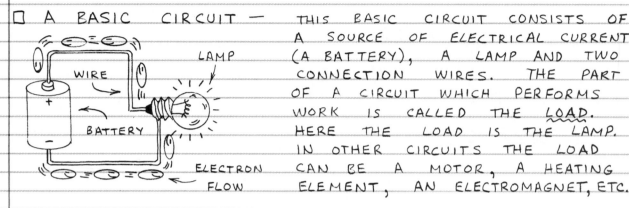

☐ A SERIES CIRCUIT — A CIRCUIT MAY INCLUDE MORE THAN ONE COMPONENT (SWITCH, LAMP, MOTOR, ETC.). A SERIES CIRCUIT IS FORMED WHEN CURRENT FLOWING THROUGH ONE COMPONENT FIRST FLOWS THROUGH ANOTHER. (ARROWS SHOW DIRECTION OF ELECTRON FLOW.)

☐ A PARALLEL CIRCUIT — A PARALLEL CIRCUIT IS FORMED WHEN TWO OR MORE COMPONENTS ARE CONNECTED SO CURRENT CAN FLOW THROUGH ONE COMPONENT WITHOUT HAVING FIRST TO FLOW THROUGH ANOTHER.

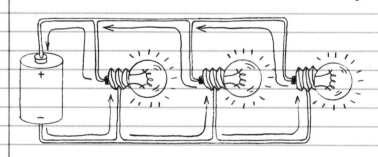

☐ A SERIES-PARALLEL CIRCUIT — MANY ELECTRICAL CIRCUITS ARE BOTH SERIES AND PARALLEL. ALL PROVIDE A COMPLETE PATH BETWEEN THE CIRCUIT AND ITS POWER SUPPLY.

☐ CIRCUIT DIAGRAMS — THUS FAR THE ELECTRICAL CIRCUITS SHOWN IN THIS BOOK HAVE BEEN ILLUSTRATED IN PICTORIAL FORM. PICTORIAL VERSIONS OF CIRCUITS WILL BE USED IN THE NEXT SEVERAL CHAPTERS AS WELL. LATER IN THE BOOK THE PICTORIALS WILL BE REPLACED BY CIRCUIT DIAGRAMS. IN A CIRCUIT DIAGRAM PICTORIAL VIEWS OF COMPONENTS ARE REPLACED BY COMPONENT SYMBOLS.

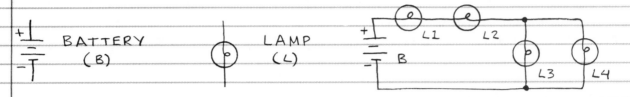

COMPONENT SYMBOLS   SERIES—PARALLEL CIRCUIT

☐ ELECTRICAL "SHORT" CIRCUIT — WHEN A WIRE OR OTHER CONDUCTOR IS PLACED ACROSS THE CONNECTIONS OF A COMPONENT, SOME OR ALL OF ANY CURRENT IN THE CIRCUIT MAY TAKE A SHORTCUT THROUGH THE CONDUCTOR. "SHORT" CIRCUITS SUCH AS THIS ARE USUALLY UNDESIRABLE AT BEST. THEY CAN CAUSE BATTERIES TO RAPIDLY LOSE THEIR CAPACITY. AND THEY CAN CAUSE DAMAGE TO WIRING AND COMPONENTS. "SHORT" CIRCUITS CAN EVEN CAUSE ENOUGH HEAT TO IGNITE THE INSULATION ON A WIRE! CAUTION: THE HUMAN BODY CONDUCTS ELECTRICITY. THEREFORE CARELESSLY TOUCHING AN ELECTRICAL CIRCUIT MAY CAUSE A "SHORT" CIRCUIT. IF THE VOLTAGE AND CURRENT ARE HIGH ENOUGH, YOU MAY RECEIVE A DANGEROUS OR EVEN LETHAL SHOCK.

☐ ELECTRICAL "GROUND" — ONE OF THE WIRES OF THE AC LINE IS CONNECTED TO EARTH BY A METAL ROD. METAL ENCLOSURES OF ELECTRICALLY POWERED DEVICES ARE CONNECTED TO THIS GROUND WIRE. THIS PREVENTS A SHOCK HAZARD SHOULD A NON-GROUNDED WIRE MAKE CONTACT WITH THE METAL ENCLOSURE. WITHOUT THE GROUND CONNECTION, A PERSON TOUCHING THE DEVICE WHILE STANDING ON THE GROUND OR A WET FLOOR MIGHT RECEIVE A DANGEROUS SHOCK. GROUND ALSO REFERS TO THE POINT IN A CIRCUIT AT ZERO VOLTAGE, WHETHER OR NOT IT'S CONNECTED TO GROUND. FOR INSTANCE, THE MINUS (−) SIDE OF THE BATTERY IN THE CIRCUITS ABOVE AND ON THE PRECEEDING PAGE CAN BE CONSIDERED GROUND.

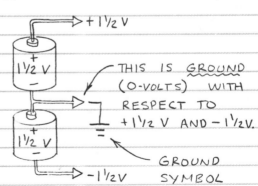

PULSES, WAVES, SIGNALS AND NOISE

ELECTRONICS IS THE STUDY AND APPLICATION OF ELECTRONS, THEIR BEHAVIOUR AND THEIR EFFECTS. THE SIMPLEST APPLICATIONS FOR ELECTRONS ARE STRAIGHTFORWARD AC AND DC CIRCUITS IN WHICH A CURRENT IS USED TO POWER LAMPS, ELECTROMAGNETS, MOTORS, SOLENOIDS AND SIMILAR DEVICES. WHAT TAKES ELECTRONICS FAR BEYOND THESE BASIC APPLICATIONS IS THE EASE WITH WHICH STREAMS OF ELECTRONS CAN BE CONTROLLED AND MANIPULATED.

THIS SIMPLE CIRCUIT IS REALLY MORE USEFUL THAN IT FIRST APPEARS BECAUSE IT CAN SEND INFORMATION BY CONVERTING A PLANNED SEQUENCE OF SWITCH CLOSURES INTO FLASHES OF LIGHT.

THE FLASHES OF THE LAMP CAN BE REPRESENTED BY A DIAGRAM LIKE THIS...

PATTERNS OF FLASHES OR PULSES LIKE THESE CAN REPRESENT COMPLEX INFORMATION LIKE SPEECH. OR SPEECH CAN BE TRANSFORMED INTO PROPORTIONAL VARIATIONS IN THE BRIGHTNESS OF A LAMP. HERE'S A SIMPLE WAY TO SEND VOICE OVER A BEAM OF REFLECTED LIGHT:

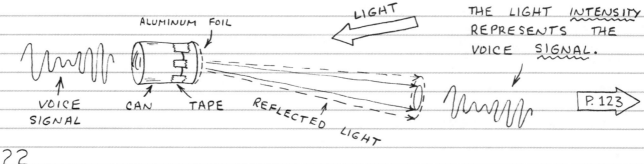

THE LIGHT INTENSITY REPRESENTS THE VOICE SIGNAL.

P. 123

☐ **PULSES** — A PULSE IS A SUDDEN, BRIEF INCREASE OR DECREASE IN A CURRENT FLOW. THE IDEAL PULSE WOULD HAVE AN INSTANTANEOUS RISE AND FALL, BUT REAL PULSES ARE NOT SO IDEAL.

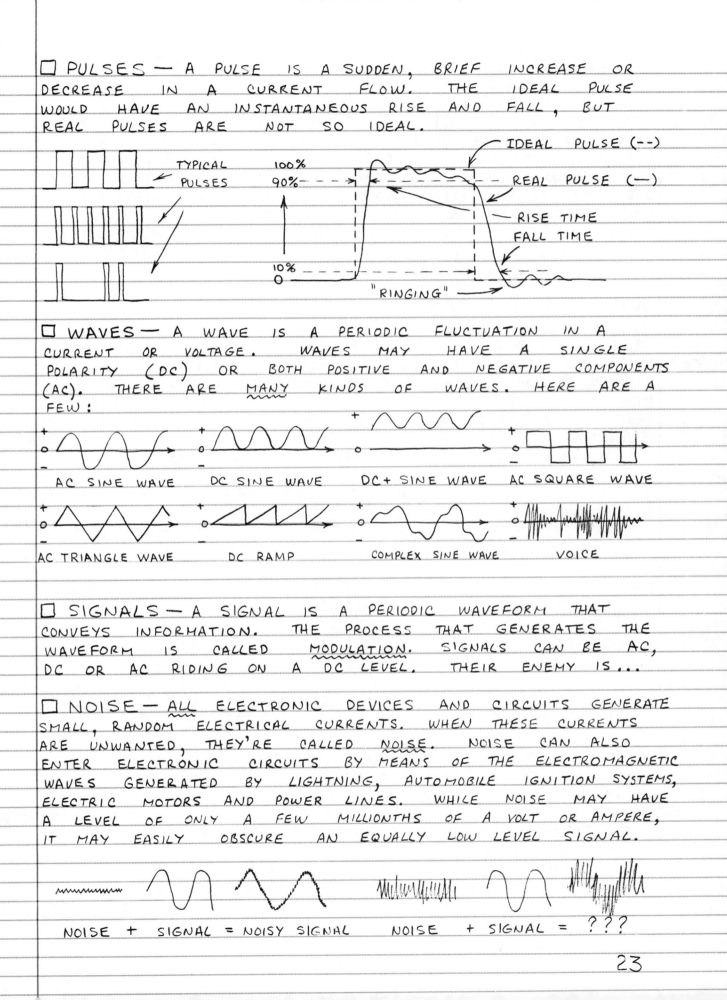

☐ **WAVES** — A WAVE IS A PERIODIC FLUCTUATION IN A CURRENT OR VOLTAGE. WAVES MAY HAVE A SINGLE POLARITY (DC) OR BOTH POSITIVE AND NEGATIVE COMPONENTS (AC). THERE ARE MANY KINDS OF WAVES. HERE ARE A FEW:

☐ **SIGNALS** — A SIGNAL IS A PERIODIC WAVEFORM THAT CONVEYS INFORMATION. THE PROCESS THAT GENERATES THE WAVEFORM IS CALLED MODULATION. SIGNALS CAN BE AC, DC OR AC RIDING ON A DC LEVEL. THEIR ENEMY IS...

☐ **NOISE** — ALL ELECTRONIC DEVICES AND CIRCUITS GENERATE SMALL, RANDOM ELECTRICAL CURRENTS. WHEN THESE CURRENTS ARE UNWANTED, THEY'RE CALLED NOISE. NOISE CAN ALSO ENTER ELECTRONIC CIRCUITS BY MEANS OF THE ELECTROMAGNETIC WAVES GENERATED BY LIGHTNING, AUTOMOBILE IGNITION SYSTEMS, ELECTRIC MOTORS AND POWER LINES. WHILE NOISE MAY HAVE A LEVEL OF ONLY A FEW MILLIONTHS OF A VOLT OR AMPERE, IT MAY EASILY OBSCURE AN EQUALLY LOW LEVEL SIGNAL.

2. ELECTRONIC COMPONENTS

Dozens of different families of parts and components block, carry, control, select, steer, switch, store, manipulate, replicate, modulate and exploit an electrical current. Those that use semiconducting crystals are so important we'll devote an entire chapter to them. You'll find just about all the remaining parts you should know about in this chapter.

WIRE AND CABLE

Used to carry an electrical current. Most wire is made from a low resistance metal like copper. Solid wire is a single conductor. Stranded wire is two or more twisted or braided bare conductors. Most wire is protected by an insulating covering of plastic, rubber or lacquer. Wire which has been tinned is easier to solder.

SPECIFICATIONS FOR BARE COPPER WIRE

GAUGE	DIAMETER (INCHES)	FEET/POUND	FEET/OHM
16	.05082	127.9	249.00
18	.04030	203.4	156.50
20	.03196	323.4	98.50
22	.02535	514.2	61.96
24	.02010	817.7	38.96
26	.01594	1300.0	24.50
28	.01264	2067.0	15.41
30	.01003	3287.0	9.69

Cables have one or more conductors and more insulation than ordinary wire. Coaxial cable can carry high frequency signals (like television).

□ CAUTION! Always use wire rated for the current it is to carry. If a wire is hot to the touch, it's carrying too much current. Use a heavier gauge wire or reduce the current. Otherwise...

SWITCHES

MECHANICAL SWITCHES PERMIT OR INTERRUPT THE FLOW OF CURRENT. THEY ARE ALSO USED TO DIRECT CURRENT TO VARIOUS POINTS.

☐ THE BASIC KNIFE SWITCH — THE SIMPLEST SWITCH...

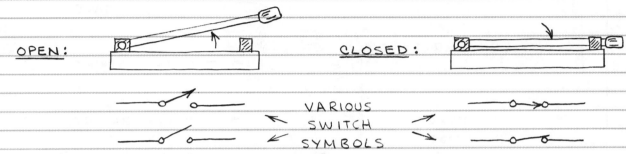

THIS IS CALLED AN SPST (SINGLE-POLE, SINGLE-THROW) SWITCH.

☐ MULTIPLE CONTACT SWITCHES — HERE ARE SYMBOLS FOR THE MAJOR KINDS:

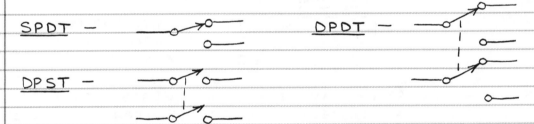

(THE DASHED LINE MEANS BOTH SIDES MOVE TOGETHER.)

SPDT — SINGLE-POLE, DOUBLE-THROW
DPST — DOUBLE-POLE, SINGLE-THROW
DPDT — DOUBLE-POLE, DOUBLE-THROW

☐ OTHER SWITCHES —

PUSHBUTTON. USUALLY SPST, NORMALLY OPEN (NO) OR NORMALLY CLOSED (NC).

NO — ◦◦ ← SPRING LOADED → NC — ◦|◦

ROTARY. WAFER-LIKE WITH ONE POLE AND 2 OR MORE CONTACTS. WAFERS CAN BE STACKED TO PROVIDE MORE POLES. MANY VARIATIONS ARE POSSIBLE.

MERCURY. MERCURY BLOB CLOSES SWITCH. POSITION SENSITIVE.

OTHER. MANY KINDS OF TOGGLE, ROCKER, LEVER, SLIDE, PUSH-ON/PUSH-OFF, ILLUMINATED AND OTHER SWITCHES ARE AVAILABLE.

RELAYS

A RELAY IS AN ELECTROMAGNETIC SWITCH. A SMALL CURRENT FLOWING THROUGH A COIL IN THE RELAY CREATES A MAGNETIC FIELD THAT PULLS ONE SWITCH CONTACT AGAINST OR AWAY FROM ANOTHER.

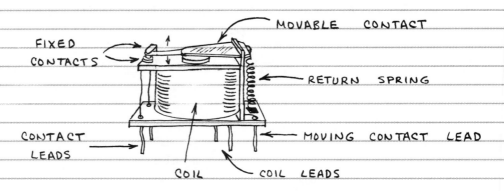

☐ RELAY SYMBOL — THE ARRANGEMENT OF CONTACTS CAN PROVIDE SPST, SPDT, DPST, DPDT AND OTHER SWITCH OPERATIONS.

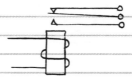

THIS IS THE SYMBOL FOR A RELAY WITH SPDT CONTACTS.

☐ REED SWITCH RELAYS — AN ENCLOSED GLASS TUBE HOUSING A PAIR OF CLOSELY SPACED SWITCH CONTACTS IS A REED SWITCH. A MAGNETIC FIELD WILL CLOSE THE CONTACTS. THIS MAKES POSSIBLE A VERY SIMPLE SPST RELAY.

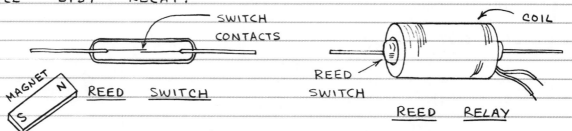

MOVING COIL METER

A COIL ON A PIVOT BETWEEN THE POLES OF A U-SHAPED MAGNET WILL ROTATE WHEN A CURRENT IS PASSED THROUGH THE COIL. THIS IS THE PRINCIPLE OF THE MOVING COIL METER.

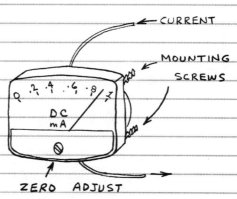

MICROPHONES AND SPEAKERS

A microphone converts sound wave variations into corresponding variations in an electrical current. The sound wave variations are first converted to back-and-forth movements of a flexible film or foil called a <u>DIAPHRAGM</u>. These movements then cause variations in an electrical current by any of the following means:

☐ CARBON — Movement of the diaphragm changes the pressure applied to a capsule of carbon particles. This causes proportional changes in the resistance of the capsule.

☐ DYNAMIC — A small coil is moved through a magnetic field as the diaphragm moves. This causes a proportional output current to be generated.

☐ CONDENSER — The moving diaphragm alters the distance between two metal plates. The result is a proportional change in the capacitance of the plates.

☐ CRYSTAL — A wafer of piezoelectric material (which produces a voltage when bent by the pressure of sound waves) forms the diaphragm or is mechanically linked to the diaphragm.

A speaker converts variations in a current or voltage into sound waves. The two most common speakers are:

☐ MAGNETIC — Similar in principle to a dynamic microphone. In fact, a magnetic speaker can be used as a microphone.

☐ CRYSTAL — Similar in principle to a crystal microphone. A crystal speaker can double as a microphone.

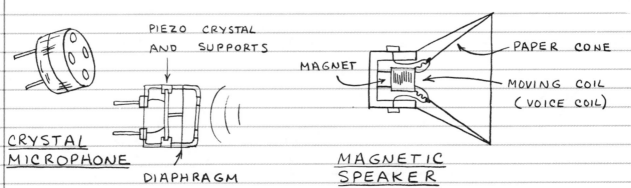

CRYSTAL MICROPHONE

MAGNETIC SPEAKER

RESISTORS

RESISTORS COME IN DOZENS OF SIZES AND SHAPES BUT THEY ALL DO THE SAME THING: <u>LIMIT</u>* <u>CURRENT</u>. MORE ABOUT THAT LATER. FIRST, LET'S SEE HOW A TYPICAL RESISTOR IS MADE:

*OR RESIST

TYPICAL CARBON COMPOSITION RESISTOR

- INCOMING CURRENT
- OUTGOING CURRENT
- WIRE LEAD
- CARBON COMPOSITION
- PROTECTIVE HOUSING
- COLOR CODE BANDS

"CARBON COMPOSITION" IS JUST A FANCY WAY OF DESCRIBING POWDERED CARBON MIXED WITH A GLUE-LIKE BINDER. THIS KIND OF RESISTOR IS EASY TO MAKE. AND ITS RESISTANCE CAN BE CHANGED FROM ONE RESISTOR TO THE NEXT SIMPLY BY CHANGING THE RATIO OF CARBON PARTICLES TO BINDER. MORE CARBON GIVES LESS RESISTANCE.

☐ DO-IT-YOURSELF RESISTORS — YOU CAN MAKE A RESISTOR BY DRAWING A LINE WITH A SOFT LEAD PENCIL ON A SHEET OF PAPER. MEASURE THE RESISTANCE OF THE LINE OR POINTS ALONG IT BY TOUCHING THE PROBES OF A MULTIMETER TO THE LINE. BE SURE TO SET THE MULTIMETER TO ITS HIGHEST RESISTANCE SCALE. THE RESISTANCE OF A SINGLE LINE MAY BE TOO HIGH TO MEASURE. IF SO, DRAW OVER THE LINE A DOZEN OR SO TIMES. HERE'S WHAT I MEASURED:

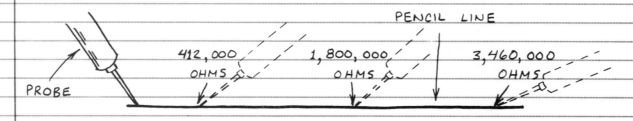

☐ RESISTOR COLOR CODE — SEE THOSE COLOR CODE BANDS ON THE RESISTOR PICTORIAL? IN REAL LIFE THEY'RE KIND OF PRETTY. BUT THEY HAVE A FAR MORE IMPORTANT PURPOSE: THEY INDICATE THE RESISTANCE OF THE RESISTOR THEY DECORATE. HERE'S HOW:

← COLOR CODE BANDS

COLOR	1	2	3 (MULTIPLIER)
BLACK	0	0	1
BROWN	1	1	10
RED	2	2	100
ORANGE	3	3	1,000
YELLOW	4	4	10,000
GREEN	5	5	100,000
BLUE	6	6	1,000,000
VIOLET	7	7	10,000,000
GRAY	8	8	100,000,000
WHITE	9	9	(NONE)

NOTE: SOMETIMES THERE'S A FOURTH BAND. IT INDICATES THE TOLERANCE* OF THE RESISTOR:

GOLD = ±5%
SILVER = ±10%
NONE = ±20%

* OR ACCURACY

LOOKS COMPLICATED THE FIRST TIME... BUT YOU'LL QUICKLY LEARN HOW TO USE IT. FOR EXAMPLE, WHAT'S THE RESISTANCE OF A RESISTOR COLOR CODED YELLOW, VIOLET AND RED? YELLOW IS THE FIRST COLOR SO THE FIRST NUMBER IS 4. VIOLET IS THE SECOND COLOR SO THE SECOND NUMBER IS 7. SINCE THE THIRD COLOR IS RED, THE MULTIPLIER IS 100. THEREFORE, THE RESISTANCE IS 47 × 100 OR 4700 OHMS. NO FOURTH COLOR BAND MEANS THE ACTUAL RESISTANCE IS 4700 ± 20%. 20% OF 4700 IS 940. THEREFORE, THE ACTUAL VALUE IS BETWEEN 3760 AND 5640 OHMS.

☐ SUBSTITUTING RESISTORS — WHAT IF YOU NEED A 6700-OHM RESISTOR BUT CAN ONLY FIND A 6800-OHM UNIT? YOU CAN ALMOST ALWAYS USE ANY VALUE WITHIN 10 OR 20% OF THE REQUIRED VALUE SO GO AHEAD AND USE IT. IF A PARTICULAR CIRCUIT REQUIRES MORE ACCURACY IT WILL TELL YOU. OF COURSE YOU CAN BUILD UP CUSTOM RESISTANCES BY CONNECTING TWO OR MORE RESISTORS IN SERIES OR IN PARALLEL. MORE ABOUT THAT LATER.

☐ **RESISTOR SUBSTITUTION PRECAUTIONS** — RESISTORS THAT CONDUCT LOTS OF CURRENT CAN BECOME VERY HOT! THEREFORE, <u>ALWAYS</u> USE RESISTORS HAVING THE PROPER POWER RATING. IF A PROJECT YOU'RE BUILDING DOESN'T SPECIFY THE POWER RATING FOR ITS RESISTORS, IT'S USUALLY OK TO USE 1/4 OR 1/2 WATT UNITS.

☐ **SOME RESISTOR SHORTHAND** — OFTEN YOU'LL SEE RESISTORS DESIGNATED WITH A K OR M SUFFIX. LIKE 47K OR 10 M. K MEANS <u>KILO</u>, AFTER THE GREEK WORD FOR 1,000. THEREFORE, 47K MEANS 47 × 1,000 OR 47,000. M IS SHORT FOR <u>MEGOHM</u> OR 1,000,000 OHMS. THEREFORE A 1M RESISTOR HAS A RESISTANCE OF 1 × 1,000,000 OR 1,000,000 OHMS. SUMMING UP...

$K = \times 1{,}000 \quad (47K = 47 \times 1{,}000 = 47{,}000 \text{ OHMS})$

$M = \times 1{,}000{,}000 \quad (2.2M = 2.2 \times 1{,}000{,}000 = 2{,}200{,}000 \text{ OHMS})$

☐ **OTHER KINDS OF RESISTORS** — THE CARBON COMPOSITION RESISTOR IS ONLY ONE OF SEVERAL MAJOR KINDS OF RESISTORS. HERE ARE OTHERS:

METAL FILM RESISTORS. VARIOUS KINDS OF RESISTORS THAT USE A THIN FILM OF METAL OR A METAL PARTICLE MIXTURE TO ACHIEVE VARIOUS RESISTANCES.

CARBON FILM RESISTORS. THESE ARE MADE BY DEPOSITING A CARBON FILM ON A SMALL CERAMIC CYLINDER. A SPIRAL GROOVE CUT INTO THE FILM CONTROLS THE LENGTH OF CARBON BETWEEN THE LEADS, HENCE THE RESISTANCE.

WIRE-WOUND RESISTORS. THESE CONSIST OF A TUBULAR FORM WRAPPED WITH COILS OF RESISTANCE WIRE. THEY ARE <u>VERY</u> ACCURATE AND CAN TAKE LOTS OF HEAT.

PHOTORESISTORS. ALSO CALLED PHOTOCELLS. MADE FROM A LIGHT SENSITIVE MATERIAL LIKE CADMIUM SULFIDE. INCREASING THE LIGHT LEVEL DECREASES THE RESISTANCE. MORE ABOUT THIS LATER.

THERMISTORS. THIS IS A TEMPERATURE SENSITIVE RESISTOR. INCREASING THE TEMPERATURE DECREASES THE RESISTANCE (IN MOST CASES).

☐ **VARIABLE RESISTORS** — OFTEN IT'S NECESSARY TO CHANGE THE RESISTANCE OF A RESISTOR. VARIABLE RESISTORS ARE CALLED <u>POTENTIOMETERS</u>. THEY ARE USED TO ALTER THE VOLUME OF A RADIO, CHANGE THE BRIGHTNESS OF A LAMP, ADJUST THE CALIBRATION OF A METER, ETC. <u>TRIMMERS</u> ARE POTENTIOMETERS EQUIPPED WITH A PLASTIC THUMBWHEEL OR A SLOT FOR A SCREWDRIVER BLADE. THEY ARE DESIGNED FOR OCCASIONAL ADJUSTMENT.

☐ **RESISTOR SYMBOLS:**

FIXED RESISTOR   POTENTIOMETER   THERMISTOR   PHOTORESISTOR

HOW RESISTORS ARE USED

☐ **SERIES CIRCUIT** — OFTEN RESISTORS ARE CONNECTED IN SERIES LIKE THIS:

THE <u>TOTAL</u> RESISTANCE IS SIMPLY THE <u>SUM</u> OF THE INDIVIDUAL RESISTANCES.

$$R_T = R1 + R2$$

☐ **PARALLEL CIRCUIT** — RESISTORS CAN ALSO BE CONNECTED IN PARALLEL LIKE THIS:

THE <u>TOTAL</u> RESISTANCE IS THE PRODUCT OF THE TWO RESISTANCES DIVIDED BY THEIR SUM.

$$R_T = \frac{R1 \times R2}{R1 + R2}$$

FOR THREE OR MORE IN PARALLEL, GO FIND YOUR CALCULATOR BECAUSE...

$$R_T = \frac{1}{\frac{1}{R1} + \frac{1}{R2} + \frac{1}{R3} \ldots \text{ETC.}}$$

☐ **VOLTAGE DIVISION** — SUPER IMPORTANT! V_{OUT} IS DETERMINED BY RATIO OF R1 AND R2. HERE'S THE FORMULA:

$$V_{OUT} = V_{IN} \left(\frac{R2}{R1 + R2} \right)$$

CAPACITORS

THERE ARE MANY KINDS OF CAPACITORS, BUT THEY ALL DO THE SAME THING: <u>STORE ELECTRONS</u>. THE SIMPLEST CAPACITOR IS TWO CONDUCTORS SEPARATED BY AN INSULATING MATERIAL CALLED THE <u>DIELECTRIC</u>. LIKE THIS:

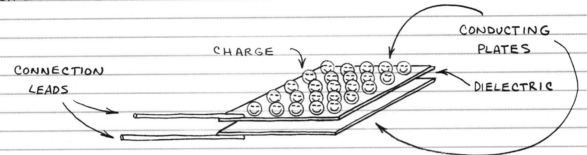

THE DIELECTRIC CAN BE PAPER, PLASTIC FILM, MICA, GLASS, CERAMIC, AIR OR A VACUUM. THE PLATES CAN BE ALUMINUM DISCS, ALUMINUM FOIL OR A THIN FILM OF METAL APPLIED TO OPPOSITE SIDES OF A SOLID DIELECTRIC. THE CONDUCTOR-DIELECTRIC-CONDUCTOR SANDWICH CAN BE ROLLED INTO A CYLINDER OR LEFT FLAT. MORE ABOUT TYPES OF CAPACITORS LATER.

HOW TO MAKE A CAPACITOR

YOU CAN MAKE A CAPACITOR FROM TWO SHEETS OF ALUMINUM FOIL AND ONE SHEET OF WAXED PAPER. FOLD THE PAPER AROUND ONE FOIL SHEET AND STACK THE SHEETS LIKE THIS:

THEN FOLD THE SHEETS LIKE THIS:

BE SURE THE FOIL SHEETS DON'T TOUCH! PRESS THE CONTACTS OF A 9-VOLT BATTERY BRIEFLY TO THE EXPOSED ENDS OF THE FOIL SHEETS. THEN TOUCH THE PROBES OF A HIGH-IMPEDANCE MULTIMETER TO THE FOIL SHEETS. THE METER WILL INDICATE A SMALL VOLTAGE FOR A FEW SECONDS. THE VOLTAGE WILL THEN FALL TO ZERO.

☐ **CHARGING A CAPACITOR** — THE MINUS SIDE OF OUR HOMEMADE CAPACITOR IS <u>CHARGED</u> WITH ELECTRONS ALMOST IMMEDIATELY. SINCE RESISTORS LIMIT CURRENT YOU CAN SLOW DOWN THE CHARGING TIME BY PLACING A RESISTOR BETWEEN THE CAPACITOR AND THE 9-VOLT BATTERY:

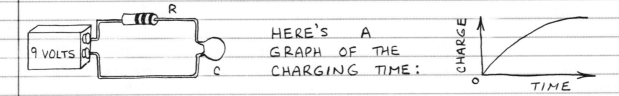

HERE'S A GRAPH OF THE CHARGING TIME:

☐ **DISCHARGING A CAPACITOR** — THE ELECTRONS IN A CHARGED CAPACITOR WILL GRADUALLY LEAK THROUGH THE DIELECTRIC UNTIL BOTH PLATES HAVE AN EQUAL CHARGE. THE CAPACITOR IS THEN <u>DISCHARGED</u>. THE CAPACITOR CAN BE DISCHARGED VERY QUICKLY BY CONNECTING ITS PLATES TOGETHER. OR IT CAN BE DISCHARGED MORE SLOWLY BY CONNECTING A RESISTOR ACROSS IT:

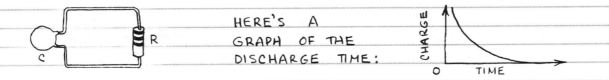

HERE'S A GRAPH OF THE DISCHARGE TIME:

☐ **SPECIFYING CAPACITORS** — THE ABILITY TO STORE ELECTRONS IS KNOWN AS <u>CAPACITANCE</u>. CAPACITANCE IS SPECIFIED IN <u>FARADS</u>. A 1-FARAD CAPACITOR CONNECTED TO A 1-VOLT SUPPLY WILL STORE 6,280,000,000,000,000,000 (6.28×10^{18}) ELECTRONS! MOST CAPACITORS HAVE MUCH SMALLER VALUES. SMALL CAPACITORS ARE SPECIFIED IN <u>PICOFARADS</u> (TRILLIONTHS OF A FARAD) AND LARGER CAPACITORS ARE SPECIFIED IN <u>MICROFARADS</u> (MILLIONTHS OF A FARAD). SUMMING UP:

1-FARAD = 1 F
1-MICROFARAD = 1 µF = 10^{-6} F = 0.000 001 F
1-PICOFARAD = 1 pF = 10^{-12} F = 0.000 000 000 001 F

☐ **SUBSTITUTING CAPACITORS** — THE CAPACITANCE SPECIFIED FOR MOST CAPACITORS MAY BE FROM 5 TO 100 % AWAY FROM THE ACTUAL VALUE. THEREFORE YOU CAN OFTEN SUBSTITUTE CLOSE VALUES FOR A SPECIFIED VALUE. BE SURE, HOWEVER, TO USE A CAPACITOR RATED AT THE EXPECTED MAXIMUM VOLTAGE LEVEL!

☐ **CAPACITOR SUBSTITUTION PRECAUTIONS** — YOU MUST MAKE SURE THE CAPACITOR YOU PLAN TO USE MEETS OR EXCEEDS THE REQUIRED VOLTAGE RATING. OTHERWISE ITS DIELECTRIC MAY BE ZAPPED BY THE STORED CHARGE. THE VOLTAGE RATING IS USUALLY PRINTED ON THE CAPACITOR. V MEANS VOLTS. WV IS WORKING VOLTS (SAME THING).

☐ **KINDS OF CAPACITORS** — CAPACITORS ARE OFTEN LABELED ACCORDING TO THEIR DIELECTRIC. THUS YOU'LL SEE REFERENCES TO CERAMIC, MICA, POLYSTYRENE AND MANY OTHERS. ALL THESE ARE <u>FIXED</u> VALUE CAPACITORS. SOME CAPACITORS HAVE A VARIABLE CAPACITY AND A SPECIAL CLASS OF FIXED CAPACITORS HAS MUCH MORE CAPACITY THAN OTHER CAPACITORS. HERE'S MORE:

VARIABLE CAPACITORS. THESE USUALLY HAVE ONE OR MORE NON-MOVING PLATES AND ONE OR MORE MOVING PLATES. THE CAPACITANCE IS CHANGED BY ROTATING A ROD AFFIXED TO ONE SIDE OF THE MOVABLE PLATES.

THIS KIND IS USED TO TUNE RADIO RECEIVERS AND TRANSMITTERS. THE DIELECTRIC IS USUALLY AIR.

THIS KIND IS USED TO TUNE OSCILLATORS LIKE THOSE USED IN DIGITAL WATCHES. THEY'RE SMALL.

ELECTROLYTIC CAPACITORS. UNIQUE IN THAT A THIN OXIDE LAYER FORMED ON ALUMINUM OR TANTALUM FOIL IS THE DIELECTRIC. MUCH HIGHER CAPACITANCE THAN NON-ELECTROLYTIC TYPES. TANTALUM UNITS HAVE MORE CAPACITANCE PER VOLUME AND A LONGER LIFE THAN ALUMINUM ELECTROLYTICS. BUT THEY COST MORE. MOST ELECTROLYTICS ARE <u>POLARIZED</u>. THEY MUST BE CONNECTED INTO A CIRCUIT IN THE PROPER DIRECTION:

 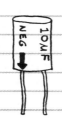 

POSITIVE LEAD <u>MUST</u> GO TO MOST POSITIVE CONNECTION POINT!

☐ **CAPACITOR SYMBOLS:**

FIXED	FIXED	VARIABLE

☐ **WARNING!** 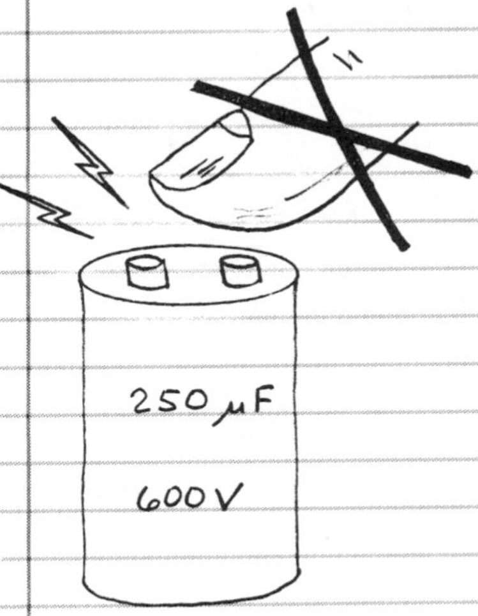 CAPACITORS CAN STORE A CHARGE FOR A CONSIDERABLE TIME AFTER THE POWER TO THEM HAS BEEN SWITCHED OFF. THIS CHARGE CAN BE <u>DANGEROUS</u>! A LARGE ELECTROLYTIC CHARGED TO ONLY 5 OR 10 VOLTS CAN MELT THE TIP OF A SCREWDRIVER PLACED ACROSS ITS TERMINALS! HIGH VOLTAGE CAPACITORS LIKE THOSE USED IN TELEVISION SETS AND PHOTOFLASH UNITS CAN STORE A <u>LETHAL</u> CHARGE! <u>NEVER</u> TOUCH THE LEADS OF SUCH A CAPACITOR. AT THE VERY LEAST THE JOLT CAN THROW YOU ACROSS A ROOM!

(250 μF, 600 V)

Sounds like the voice of experience!

HOW CAPACITORS ARE USED

☐ **PARALLEL CIRCUIT** — OFTEN CAPACITORS ARE CONNECTED IN PARALLEL LIKE THIS:

THE <u>TOTAL</u> CAPACITANCE IS THE SUM OF THE INDIVIDUAL CAPACITANCES.

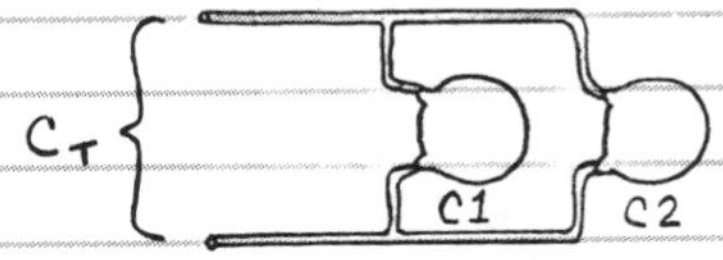

$$C_T = C_1 + C_2$$

☐ **SERIES CIRCUIT** — SOMETIMES CAPACITORS ARE CONNECTED IN SERIES LIKE THIS:

THE <u>TOTAL</u> CAPACITANCE IS THE PRODUCT OF THE TWO CAPACITANCES DIVIDED BY THEIR SUM.

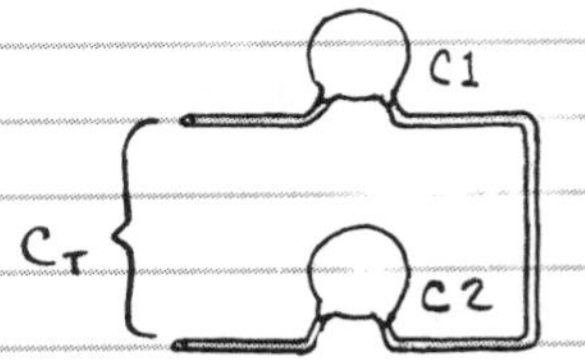

$$C_T = \frac{C_1 \times C_2}{C_1 + C_2}$$

THREE OR MORE CAPACITORS IN SERIES? HERE'S THE FORMULA:

$$C_T = \frac{1}{\frac{1}{C_1} + \frac{1}{C_2} + \frac{1}{C_3} \ldots} \text{ ETC.}$$

☐ **AND MORE** — THERE ARE MANY OTHER WAYS TO USE CAPACITORS, SOME OF WHICH ARE SHOWN NEXT...

Resistor and Capacitor Applications

Resistors and capacitors are the **key** ingredients of many electronic circuits. Here are some reasons why:

☐ **Power Supply Filter** — A capacitor will smooth (filter) the pulsating voltage from a power supply into a steady direct current (DC).

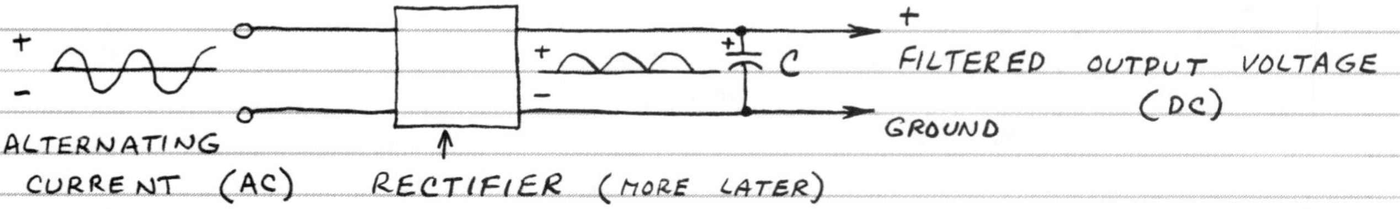

[Diagram: AC input → Rectifier (more later) → pulsating DC → Capacitor C → Filtered output voltage (DC) / Ground]

☐ **Spike Remover** — Digital logic circuits, which we'll find out more about later, can use lots of current momentarily when they switch from off to on or vice versa. This can cause very brief but substantial reductions in power applied to nearby circuits. These power **spikes** (or **glitches**, as they are sometimes called) can be eliminated by placing a small (0.1 μF) capacitor across the power leads of the logic circuit:

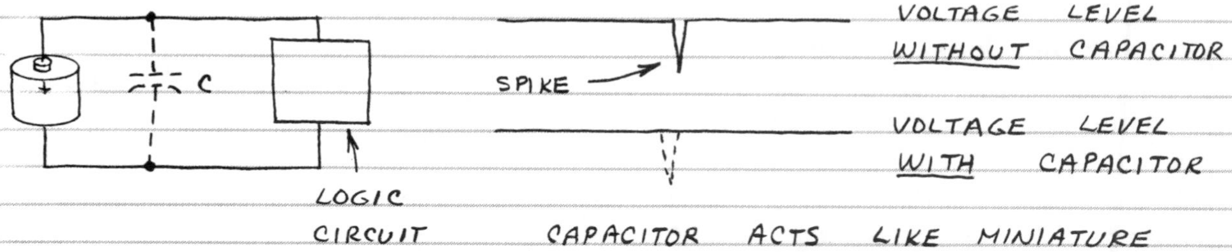

[Diagram: Battery — Capacitor C — Logic circuit; waveform showing spike (voltage level without capacitor) vs. smooth (voltage level with capacitor)]

Capacitor acts like miniature battery that supplies power during the spike.

☐ **AC-DC Selective Filter** — Often an electrical signal will be riding atop a steady DC signal. For example, the signal from a lightwave communication system may look like this when it's dark: But sunlight causes this:

[Two waveform diagrams; second shows signal with steady DC offset caused by sunlight] → P.123

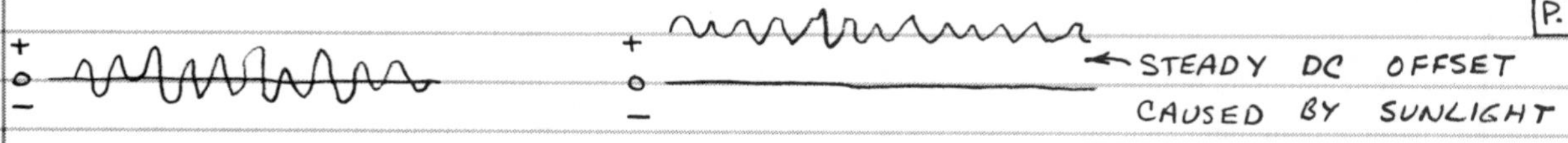

A capacitor will pass the fluctuating signal and completely block the steady DC level.

☐ **R-C CIRCUITS** — TWO CIRCUITS THAT COMBINE A RESISTOR (R) AND CAPACITOR (C) ARE <u>VERY</u> IMPORTANT. THEY ARE THE <u>INTEGRATOR</u> AND <u>DIFFERENTIATOR</u>. BOTH THESE CIRCUITS ARE USED TO RESHAPE AN INCOMING STREAM OF WAVES OR PULSES.

THE PRODUCT OF R AND C IN THESE CIRCUITS IS CALLED THE <u>RC TIME CONSTANT</u>. FOR THE CIRCUITS SHOWN BELOW, THE RC TIME CONSTANT (IN SECONDS) IS AT LEAST TEN TIMES THE INTERVAL BETWEEN INCOMING CYCLES OR PULSES.

1. INTEGRATOR. HERE'S A BASIC RC INTEGRATOR:

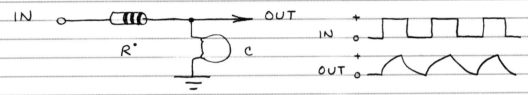

IF THE INPUT PULSES ARE SPEEDED UP, THE OUTPUT WAVEFORMS (OFTEN CALLED A SAWTOOTH) WILL NOT REACH THEIR FULL HEIGHT (AMPLITUDE). IT'S EASY TO DESIGN AN AMPLIFIER THAT IGNORES WAVES WITH LESS THAN A DESIRED AMPLITUDE. THEREFORE, THE INTEGRATOR CAN FUNCTION AS A FILTER WHICH PASSES ONLY SIGNALS BELOW A CERTAIN FREQUENCY.

2. DIFFERENTIATOR. HERE'S A BASIC RC DIFFERENTIATOR:

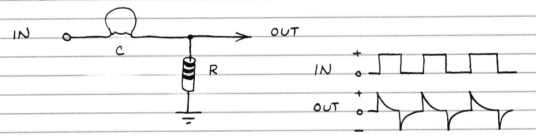

THIS CIRCUIT PRODUCES SYMMETRICAL OUTPUT WAVES WITH SHARP POSITIVE <u>AND</u> NEGATIVE PEAKS. IT'S USED TO MAKE NARROW PULSE GENERATORS FOR TELEVISION RECEIVERS AND TO TRIGGER DIGITAL LOGIC CIRCUITS.

☐ **MORE ABOUT R-C** — YOU WILL OFTEN SEE REFERENCES TO THE RC TIME CONSTANT OF A CIRCUIT. IT'S THE TIME IN SECONDS FOR A CHARGING OR DISCHARGING CAPACITOR TO GO THROUGH 63.3% OF THE CHANGE IN CHARGE.

COILS

Electrons moving through a wire cause an _electromagnetic field_ to encircle the wire. As you know from Chapter 1, passing a current through a wire that's been wrapped as a coil (p. 15) creates an even stronger field. This field makes possible solenoids, motors and electromagnets. Coils have other important roles, too:

1. Coils resist _rapid_ changes in the current flowing through them while freely passing steady (DC) current. Here are some examples:

SIGNAL	IN	COIL	OUT
SLOW SINE WAVE	∿∿	—⦙⦙⦙→	∿∿
FAST SINE WAVE	∿∿∿∿∿	—⦙⦙⦙→	∼∼
SLOW SQUARE WAVE	⊓⊔⊓	—⦙⦙⦙→	/\/\
FAST SQUARE WAVE	⊓⊔⊓⊔⊓⊔	—⦙⦙⦙→	∼∼

Sometimes a coil will add _ringing_ to a square wave passing through it. This can happen when the resistance of the external current path that connects the ends of the coil is high.

SQUARE WAVE — ⊓⊔⊓ —⦙⦙⦙→ [ringing waveform]

👉 THIS IS _RINGING_

2. Some of the energy in the field around a coil can be _induced_ (transferred) into a second, nearby coil. This is the principle of the _transformer_:

IN ∿∿ [transformer with two coils and field lines] OUT ∿∿
↑ FIELD

The input side of the transformer is called the _PRIMARY_. The output side is called the _SECONDARY_.

38

☐ **TYPES OF COILS** — THERE ARE MANY DIFFERENT TYPES OF COILS. HERE ARE SOME OF THEM:

<u>TUNING COIL</u>. RADIOS USE VARIOUS COILS TO HELP SELECT A DESIRED SIGNAL. TUNING COILS HAVE A SERIES OF TAPS OR A MOVABLE CORE SO THEIR <u>INDUCTANCE</u>*, HENCE <u>RESONANT FREQUENCY</u>, CAN BE CHANGED.

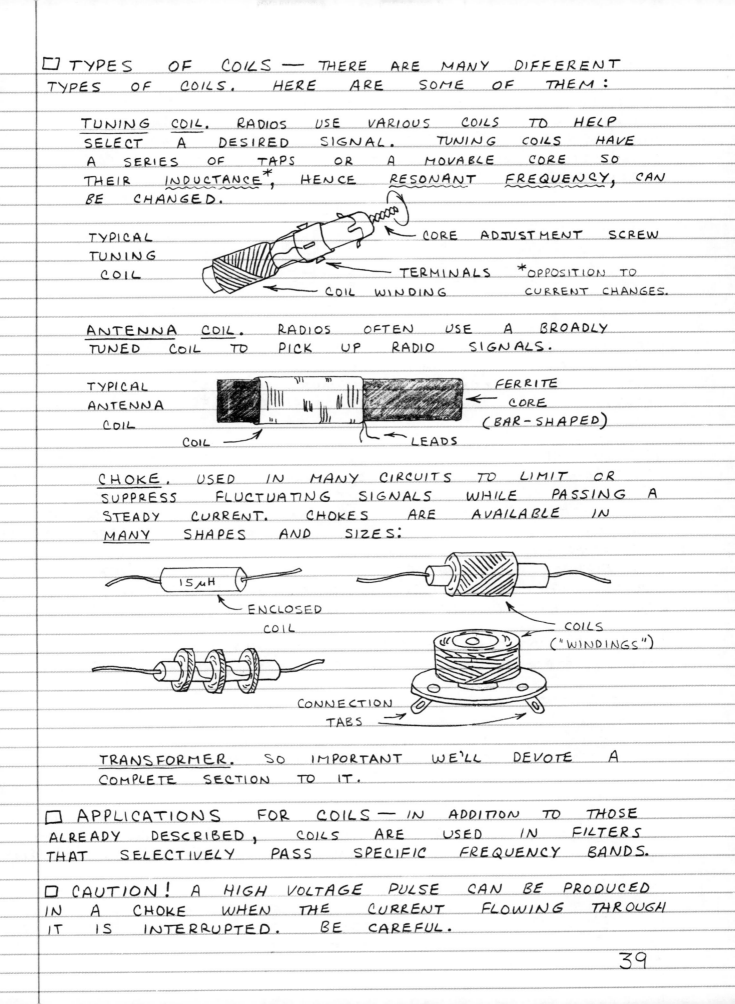

TYPICAL TUNING COIL
← CORE ADJUSTMENT SCREW
← TERMINALS
← COIL WINDING
*OPPOSITION TO CURRENT CHANGES.

<u>ANTENNA COIL</u>. RADIOS OFTEN USE A BROADLY TUNED COIL TO PICK UP RADIO SIGNALS.

TYPICAL ANTENNA COIL
← FERRITE CORE (BAR-SHAPED)
← COIL
← LEADS

<u>CHOKE</u>. USED IN MANY CIRCUITS TO LIMIT OR SUPPRESS FLUCTUATING SIGNALS WHILE PASSING A STEADY CURRENT. CHOKES ARE AVAILABLE IN MANY SHAPES AND SIZES:

15μH
← ENCLOSED COIL
← COILS ("WINDINGS")
← CONNECTION TABS

<u>TRANSFORMER</u>. SO IMPORTANT WE'LL DEVOTE A COMPLETE SECTION TO IT.

☐ **APPLICATIONS FOR COILS** — IN ADDITION TO THOSE ALREADY DESCRIBED, COILS ARE USED IN FILTERS THAT SELECTIVELY PASS SPECIFIC FREQUENCY BANDS.

☐ **CAUTION!** A HIGH VOLTAGE PULSE CAN BE PRODUCED IN A CHOKE WHEN THE CURRENT FLOWING THROUGH IT IS INTERRUPTED. BE CAREFUL.

TRANSFORMERS

Transformers are a major class of coils having two or more windings usually wrapped around a common core made from laminated iron sheets. Here's a simple transformer:

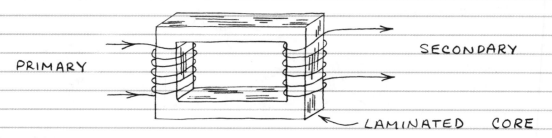

If the current flowing through the primary coil is fluctuating, then a current will be induced into the secondary winding. A steady (DC) current will <u>not</u> be transferred from one coil to the other.

☐ **HOW THEY WORK** — Transformers have the ability to transform voltage and current to higher or lower levels. They do <u>not</u>, of course, create power from nothing. Therefore, if a transformer boosts the voltage of a signal, it reduces its current. And if it cuts the voltage of a signal, it raises its current. In other words... the power flowing from a transformer <u>cannot</u> exceed the incoming power!

☐ **TURNS RATIO** — The ratio of primary to secondary turns determines a transformer's voltage ratio...

1:1 RATIO.

 The primary voltage and current are transferred unaltered to the secondary. Often called an <u>ISOLATION TRANSFORMER</u>.

STEP-UP.

 The voltage is increased by the turns ratio. Thus a 1:5 turns ratio will boost 5-volts at the primary into 25-volts at the secondary.

STEP-DOWN.

 The voltage is reduced by the turns ratio. Thus a 5:1 turns ratio will drop 25-volts at the primary to 5-volts at the secondary.

☐ TRANSFORMER TYPES AND APPLICATIONS — HERE ARE SOME OF THE MAJOR TRANSFORMER TYPES:

ISOLATION.

USED TO ISOLATE DIFFERENT PARTS OF A CIRCUIT AND TO PROVIDE PROTECTION FROM ELECTRICAL SHOCK.

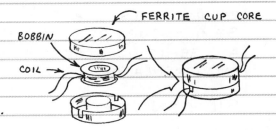

STANDARD 1:1 ISOLATION

MINIATURE 1:1 ISOLATION

POWER CONVERSION.

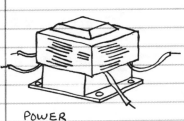

OFTEN USED TO REDUCE POWER LINE VOLTAGE TO USABLE LEVEL.

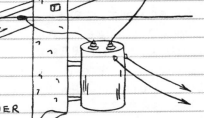

POWER TRANSFORMER

UTILITY COMPANY TRANSFORMER

HIGH-VOLTAGE.

AUTOMOTIVE IGNITION COIL

USED TO PRODUCE IGNITION SPARKS IN GASOLINE ENGINES. ALSO USED TO POWER TV PICTURE TUBES, SOME LASERS, NEON LIGHTS, ETC.

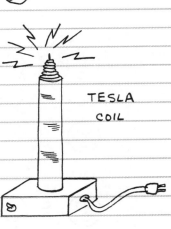

TESLA COIL

AUDIO.

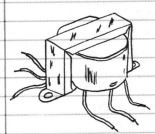

USED TO MATCH THE IMPEDANCE* OF AN AMPLIFIER TO THAT OF A MICROPHONE, SPEAKER OR OTHER DEVICE.

*OPPOSITION TO THE FLOW OF ALTERNATING CURRENT.

MINIATURE

TAPPED PRIMARY AND SECONDARY WINDINGS

NOTE: LEADS OF TRANSFORMERS ARE COLOR CODED.

3. SEMICONDUCTORS

The most exciting and important electronic components are made from crystals called SEMICONDUCTORS. Depending on certain conditions, a semiconductor can act like a conductor or an insulator.

SILICON

There are many different semiconducting materials, but silicon, the main ingredient of sand, is the most popular.

A silicon atom has but four electrons in its outermost shell, but it would like to have eight. Therefore, a silicon atom will link up with four of its neighbors to share electrons:

SILICON ATOM — ELECTRONS, NUCLEUS, SHELLS

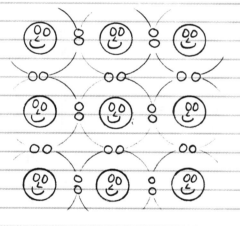

A cluster of silicon atoms sharing outer electrons forms a regular arrangement called a CRYSTAL.

← This is a magnified view of a silicon crystal. To keep things simple, only the outer electrons of each atom are shown.

Silicon forms 27.7% of the Earth's crust! Only oxygen is more common. It's never found in the pure state. When purified, it's dark gray in color.

Silicon and diamond share the same crystal structure and other properties. But silicon is not transparent.

"SEED" CRYSTAL — BOULE — MOLTEN SILICON — 2,570°F

Silicon can be grown into big crystals. It's cut into wafers for making electronic parts.

42

☐ **SILICON RECIPES** — PURE SILICON ISN'T VERY USEFUL. THAT'S WHY SILICON MAKERS SPICE UP THEIR SILICON RECIPES WITH A DASH OF PHOSPHORUS, BORON OR OTHER GOODIES. THIS IS CALLED DOPING THE SILICON. WHEN GROWN INTO CRYSTALS, DOPED SILICON HAS VERY USEFUL ELECTRONIC PROPERTIES!

☐ **P & N SPICED SILICON LOAF** — BORON, PHOSPHORUS AND CERTAIN OTHER ATOMS CAN JOIN WITH SILICON ATOMS TO FORM CRYSTALS. HERS'S THE CATCH: A BORON ATOM HAS ONLY THREE ELECTRON IN ITS OUTER SHELL. AND A PHOSPHORUS ATOM HAS FIVE ELECTRONS IN ITS OUTER SHELL. SILICON WITH EXTRA PHOSPHORUS ELECTRONS IS CALLED N-TYPE SILICON (N = NEGATIVE). SILICON WITH ELECTRON DEFICIENT BORON ATOMS IS CALLED P-TYPE SILICON (P = POSITIVE).

☐ **P-TYPE SILICON** —
A BORON ATOM IN A CLUSTER OF SILICON ATOMS LEAVES A VACANT ELECTRON OPENING CALLED A HOLE. IT'S POSSIBLE FOR AN ELECTRON FROM A NEARBY ATOM TO "FALL" INTO THE HOLE. THEREFORE, THE HOLE HAS MOVED TO A NEW LOCATION. REMEMBER, HOLES CAN MOVE THROUGH SILICON (JUST AS BUBBLES MOVE THROUGH WATER).

☐ **N-TYPE SILICON** —
A PHOSPHORUS ATOM IN A CLUSTER OF SILICON ATOMS DONATES AN EXTRA ELECTRON. THIS EXTRA ELECTRON CAN MOVE THROUGH THE CRYSTAL WITH COMPARATIVE EASE. IN OTHER WORDS, N-TYPE SILICON CAN CARRY AN ELECTRICAL CURRENT. BUT SO CAN P-TYPE SILICON! HOLES "CARRY" THE CURRENT.

THE DIODE

BOTH P-TYPE AND N-TYPE SILICON CONDUCT ELECTRICITY. THE RESISTANCE OF BOTH TYPES IS DETERMINED BY THE PROPORTION OF HOLES OR SURPLUS ELECTRONS. THEREFORE BOTH TYPES CAN FUNCTION AS RESISTORS. AND THEY WILL CONDUCT ELECTRICITY IN ANY DIRECTION.

BY FORMING SOME P-TYPE SILICON IN A CHIP OF N-TYPE SILICON, ELECTRONS WILL FLOW THROUGH THE SILICON IN ONLY ONE DIRECTION. THIS IS THE PRINCIPLE OF THE DIODE. THE P-N INTERFACE IS CALLED THE PN JUNCTION.

☐ HOW THE DIODE WORKS — HERE'S A SIMPLIFIED EXPLANATION OF HOW A DIODE CONDUCTS ELECTRICITY IN ONE DIRECTION (FORWARD) WHILE BLOCKING THE FLOW OF CURRENT IN THE OPPOSITE DIRECTION (REVERSE).

FORWARD BIAS	REVERSE BIAS

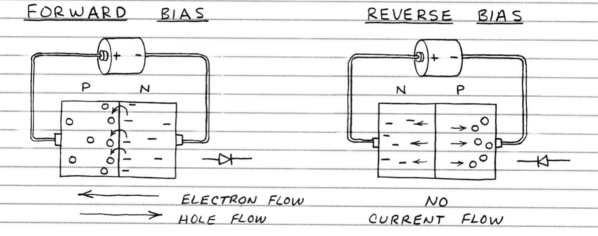

| ← ELECTRON FLOW | NO |
| → HOLE FLOW | CURRENT FLOW |

HERE THE CHARGE FROM THE BATTERY REPELS HOLES AND ELECTRONS TOWARD THE JUNCTION. IF THE VOLTAGE EXCEEDS 0.6-VOLT (SILICON), THEN ELECTRONS WILL CROSS THE JUNCTION AND COMBINE WITH HOLES. A CURRENT THEN FLOWS.

HERE THE CHARGE FROM THE BATTERY ATTRACTS HOLES AND ELECTRONS AWAY FROM THE JUNCTION. THEREFORE, NO CURRENT CAN FLOW.

☐ A TYPICAL DIODE — DIODES ARE COMMONLY ENCLOSED IN SMALL GLASS CYLINDERS. A DARK BAND MARKS THE CATHODE TERMINAL. THE OPPOSITE TERMINAL IS THE ANODE.

CATHODE
SYMBOL
CURRENT FLOWS WHEN ANODE IS MORE POSITIVE THAN CATHODE.
ANODE

☐ DIODE OPERATION— YOU ALREADY KNOW A DIODE IS LIKE AN ELECTRONIC ONE-WAY VALVE. IT'S IMPORTANT TO UNDERSTAND SOME ADDITIONAL ASPECTS OF DIODE OPERATION. HERE ARE SOME KEY ONES:

1. A DIODE WILL NOT CONDUCT UNTIL THE FORWARD VOLTAGE REACHES A CERTAIN THRESHOLD POINT. FOR SILICON DIODES THIS VOLTAGE IS ABOUT 0.6-VOLT.

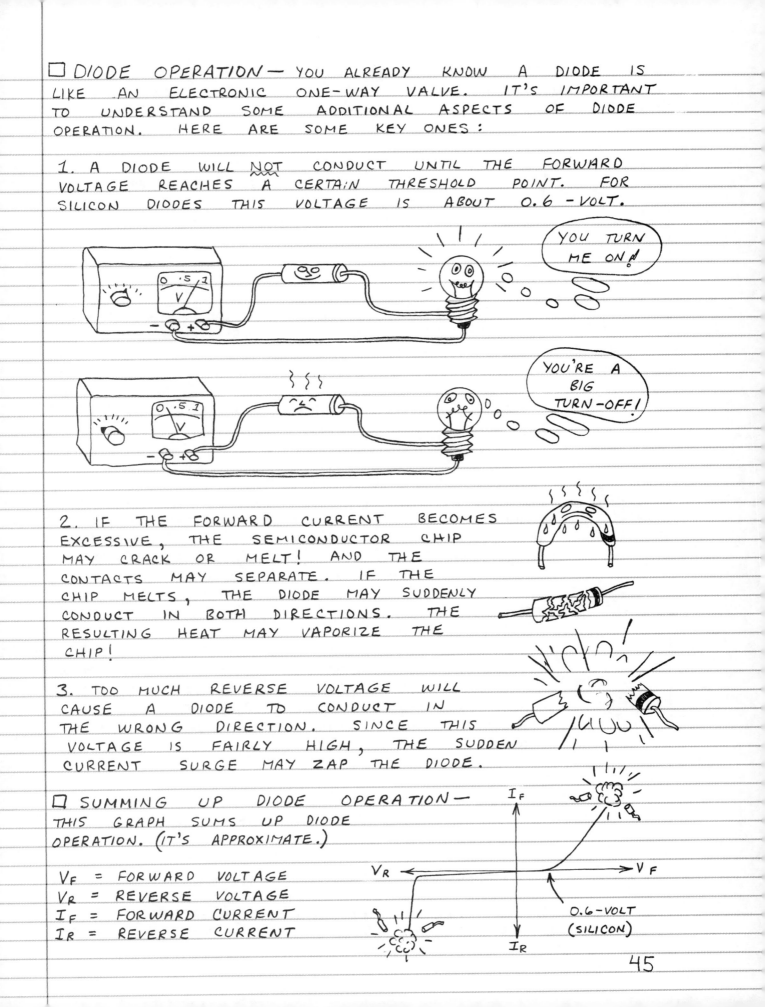

2. IF THE FORWARD CURRENT BECOMES EXCESSIVE, THE SEMICONDUCTOR CHIP MAY CRACK OR MELT! AND THE CONTACTS MAY SEPARATE. IF THE CHIP MELTS, THE DIODE MAY SUDDENLY CONDUCT IN BOTH DIRECTIONS. THE RESULTING HEAT MAY VAPORIZE THE CHIP!

3. TOO MUCH REVERSE VOLTAGE WILL CAUSE A DIODE TO CONDUCT IN THE WRONG DIRECTION. SINCE THIS VOLTAGE IS FAIRLY HIGH, THE SUDDEN CURRENT SURGE MAY ZAP THE DIODE.

☐ SUMMING UP DIODE OPERATION— THIS GRAPH SUMS UP DIODE OPERATION. (IT'S APPROXIMATE.)

V_F = FORWARD VOLTAGE
V_R = REVERSE VOLTAGE
I_F = FORWARD CURRENT
I_R = REVERSE CURRENT

◻ TYPES OF DIODES — MANY DIFFERENT KINDS OF DIODES ARE AVAILABLE. HERE ARE SOME OF THE MAJOR TYPES:

SMALL SIGNAL.

SMALL SIGNAL DIODES ARE USED TO TRANSFORM LOW CURRENT AC TO DC, DETECT (DEMODULATE) RADIO SIGNALS, MULTIPLY VOLTAGE, PERFORM LOGIC, ABSORB VOLTAGE SPIKES, ETC.

POWER RECTIFIER.

FUNCTIONALLY IDENTICAL TO SMALL SIGNAL DIODES, POWER RECTIFIERS CAN HANDLE MUCH MORE CURRENT. THEY ARE INSTALLED IN LARGE METAL PACKAGES THAT SOAK UP EXCESS HEAT AND TRANSFER IT TO A METAL HEAT SINK. USED MAINLY IN POWER SUPPLIES.

ZENER.

THE ZENER DIODE IS DESIGNED TO HAVE A SPECIFIC <u>REVERSE BREAKDOWN</u> (CONDUCTION) <u>VOLTAGE</u>. THIS MEANS ZENER DIODES CAN FUNCTION LIKE A VOLTAGE SENSITIVE SWITCH. ZENER DIODES HAVING BREAKDOWN VOLTAGES (V_z) OF FROM ABOUT 2-VOLTS TO 200-VOLTS ARE AVAILABLE.

LIGHT-EMITTING.

ALL DIODES EMIT SOME ELECTROMAGNETIC RADIATION WHEN FORWARD BIASED. DIODES MADE FROM CERTAIN SEMICONDUCTORS (LIKE GALLIUM ARSENIDE PHOSPHIDE) EMIT <u>CONSIDERABLY</u> MORE RADIATION THAN SILICON DIODES. THEY'RE CALLED LIGHT-EMITTING DIODES (LEDs).

PHOTODIODE.

ALL DIODES RESPOND TO SOME DEGREE WHEN ILLUMINATED BY LIGHT. DIODES DESIGNED SPECIFICALLY TO DETECT LIGHT ARE CALLED PHOTODIODES. THEY INCLUDE A GLASS OR PLASTIC WINDOW THROUGH WHICH THE LIGHT ENTERS. OFTEN THEY HAVE A LARGE, EXPOSED JUNCTION REGION. SILICON MAKES GOOD PHOTODIODES.

HOW DIODES ARE USED

IN CHAPTER 9 YOU'LL SEE HOW VARIOUS TYPES OF DIODES ARE USED IN MANY APPLICATIONS. FOR NOW HERE ARE TWO OF THE MOST IMPORTANT ROLES FOR SMALL SIGNAL DIODES AND RECTIFIERS:

☐ HALF-WAVE RECTIFIER

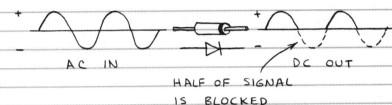

AN UNDULATING (AC) SIGNAL (OR VOLTAGE) IS RECTIFIED INTO A SINGLE POLARITY (DC) SIGNAL (OR VOLTAGE).

☐ FULL-WAVE RECTIFIER

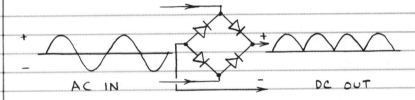

THIS 4-DIODE "NETWORK" (OR BRIDGE RECTIFIER) RECTIFIES BOTH HALVES OF AN AC SIGNAL.

MORE ABOUT THE DIRECTION OF CURRENT FLOW

AN ELECTRICAL CURRENT IS THE MOVEMENT OF ELECTRONS THROUGH A CONDUCTOR OR SEMICONDUCTOR. SINCE ELECTRONS MOVE FROM A NEGATIVELY CHARGED TO A POSITIVELY CHARGED REGION, WHY DOES THE ARROWHEAD IN A DIODE SYMBOL POINT IN THE OPPOSITE DIRECTION? THERE ARE TWO REASONS:

1. BEGINNING WITH BENJAMIN FRANKLIN, IT WAS TRADITIONALLY ASSUMED ELECTRICITY FLOWS FROM A POSITIVELY CHARGED TO A NEGATIVELY CHARGED REGION. THE DISCOVERY OF THE ELECTRON CORRECTED THAT. (BUT MOST ELECTRICAL CIRCUIT DIAGRAMS TODAY STILL FOLLOW THE OLD TRADITION IN WHICH THE POSITIVE POWER SUPPLY CONNECTION IS PLACED ABOVE THE NEGATIVE CONNECTION AS IF GRAVITY SOMEHOW INFLUENCES THE FLOW OF A CURRENT.)

2. IN A SEMICONDUCTOR, AS SHOWN ON PAGE 44, HOLES FLOW IN THE DIRECTION OPPOSITE THAT OF ELECTRON FLOW. IT'S THEREFORE COMMON TO REFER TO POSITIVE CURRENT FLOW IN SEMICONDUCTORS.

FOR ACCURACY, IN THIS BOOK "CURRENT FLOW" REFERS TO ELECTRON FLOW. BUT WE'RE STUCK WITH SYMBOLS THAT INDICATE HOLE FLOW.

THE TRANSISTOR

TRANSISTORS ARE SEMICON-DUCTOR DEVICES WITH THREE LEADS. A VERY SMALL CURRENT OR VOLTAGE AT ONE LEAD CAN CONTROL A MUCH LARGER CURRENT FLOWING THROUGH THE OTHER TWO LEADS. THIS MEANS TRANSISTORS CAN BE USED AS AMPLIFIERS AND SWITCHES. THERE ARE TWO MAIN FAMILIES OF TRANSISTORS: BIPOLAR AND FIELD-EFFECT.

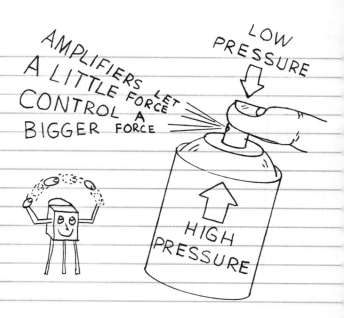

BIPOLAR TRANSISTORS

ADD A SECOND JUNCTION TO A PN JUNCTION DIODE AND YOU GET A 3-LAYER SILICON SANDWICH. THE SANDWICH CAN BE EITHER NPN OR PNP. EITHER WAY, THE MIDDLE LAYER ACTS LIKE A FAUCET OR GATE THAT CONTROLS THE CURRENT MOVING THROUGH THE THREE LAYERS.

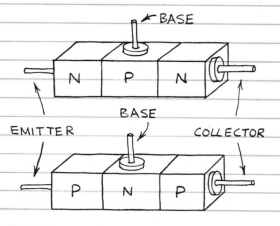

☐ BIPOLAR TRANSISTOR OPERATION — THE THREE LAYERS OF A BIPOLAR TRANSISTOR ARE THE EMITTER, BASE AND COLLECTOR. THE BASE IS VERY THIN AND HAS FEWER DOPING ATOMS THAN THE EMITTER AND COLLECTOR. THEREFORE A VERY SMALL EMITTER-BASE CURRENT WILL CAUSE A MUCH LARGER EMITTER-COLLECTOR CURRENT TO FLOW.

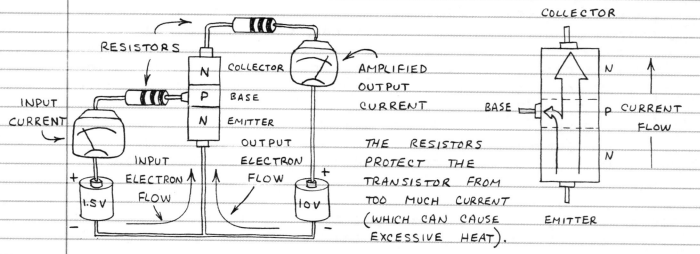

THE RESISTORS PROTECT THE TRANSISTOR FROM TOO MUCH CURRENT (WHICH CAN CAUSE EXCESSIVE HEAT).

☐ MORE ABOUT BIPOLAR TRANSISTOR OPERATION — DIODES AND TRANSISTORS SHARE SEVERAL KEY FEATURES:

1. THE BASE-EMITTER JUNCTION (OR DIODE) WILL NOT CONDUCT UNTIL THE FORWARD VOLTAGE EXCEEDS 0.6-VOLT.

2. TOO MUCH CURRENT WILL CAUSE A TRANSISTOR TO BECOME HOT AND OPERATE IMPROPERLY. IF A TRANSISTOR IS HOT WHEN TOUCHED, DISCONNECT THE POWER TO IT!

3. TOO MUCH CURRENT OR VOLTAGE MAY DAMAGE OR PERMANENTLY DESTROY THE SEMICONDUCTOR CHIP THAT FORMS A TRANSISTOR. IF THE CHIP ISN'T HARMED, ITS TINY CONNECTION WIRES MAY MELT OR SEPARATE FROM THE CHIP. NEVER CONNECT A TRANSISTOR BACKWARDS!

☐ KINDS OF TRANSISTORS — MANY DIFFERENT KINDS OF TRANSISTORS ARE AVAILABLE. HERE ARE EXAMPLES OF THE MOST IMPORTANT:

SMALL SIGNAL AND SWITCHING.

SMALL SIGNAL TRANSISTORS ARE USED TO AMPLIFY LOW LEVEL SIGNALS. SWITCHING TRANSISTORS ARE DESIGNED TO BE OPERATED FULLY ON OR OFF. SOME TRANSISTORS CAN BOTH AMPLIFY AND SWITCH EQUALLY WELL.

POWER.

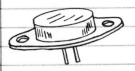

POWER TRANSISTORS ARE USED IN HIGH POWER AMPLIFIERS AND POWER SUPPLIES. LARGE SIZE AND EXPOSED METAL SURFACES KEEP THEM COOL.

HIGH-FREQUENCY.

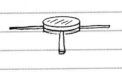

HIGH-FREQUENCY TRANSISTORS OPERATE AT RADIO, TELEVISION AND MICROWAVE FREQUENCIES. THE BASE REGION IS VERY THIN AND THE ACTUAL CHIP IS VERY SMALL.

☐ BIPOLAR TRANSISTOR SYMBOLS — ARROWS POINT IN DIRECTION OF HOLE FLOW.

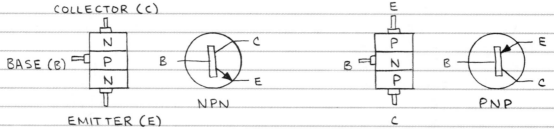

HOW BIPOLAR TRANSISTORS ARE USED

WHEN THE BASE OF AN NPN TRANSISTOR IS GROUNDED (0 VOLTS), NO CURRENT FLOWS FROM THE EMITTER TO THE COLLECTOR (THE TRANSISTOR IS "OFF"). IF THE BASE IS FORWARD-BIASED BY AT LEAST 0.6 VOLT, A CURRENT WILL FLOW FROM THE EMITTER TO THE COLLECTOR (THE TRANSISTOR IS "ON"). WHEN OPERATED IN ONLY THESE TWO MODES, THE TRANSISTOR FUNCTIONS AS A SWITCH. IF THE BASE IS FORWARD-BIASED, THE EMITTER-COLLECTOR CURRENT WILL FOLLOW VARIATIONS IN A MUCH SMALLER BASE CURRENT. THE TRANSISTOR THEN FUNCTIONS AS AN AMPLIFIER. THIS DISCUSSION APPLIES TO A TRANSISTOR IN WHICH THE EMITTER IS THE GROUND CONNECTION FOR BOTH THE INPUT AND OUTPUT AND IS CALLED THE COMMON-EMITTER CIRCUIT. SOME SIMPLIFIED COMMON-EMITTER CIRCUITS ARE SHOWN BELOW. SO YOU CAN SEE HOW THEY ARE USED IN REAL CIRCUITS, EACH EXAMPLE REFERS TO A TYPICAL WORKING APPLICATION IN CHAPTER 9.

➡ P. 92

☐ A BIPOLAR TRANSISTOR SWITCH

➡ P. 107

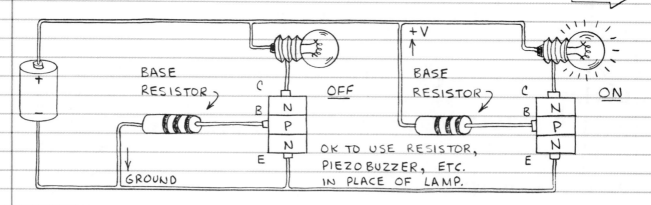

ONLY TWO INPUTS ARE POSSIBLE: GROUND (0 VOLTS) AND THE POSITIVE BATTERY VOLTAGE (+V). THEREFORE THE TRANSISTOR IS OFF OR ON. A TYPICAL BASE RESISTANCE IS 5,000 TO 10,000 OHMS. (IF THE RESISTOR IS REPLACED BY A WIRE, THE LAMP CAN BE SWITCHED ON OR OFF FROM A CONSIDERABLE DISTANCE.)

☐ **A BIPOLAR TRANSISTOR DC AMPLIFIER** — THE VARIABLE RESISTOR FORWARD BIASES THE TRANSISTOR AND CONTROLS THE INPUT (BASE-EMITTER) CURRENT. THE METER INDICATES THE OUTPUT (COLLECTOR-EMITTER) CURRENT. THE SERIES RESISTOR PROTECTS THE METER FROM EXCESSIVE CURRENT.

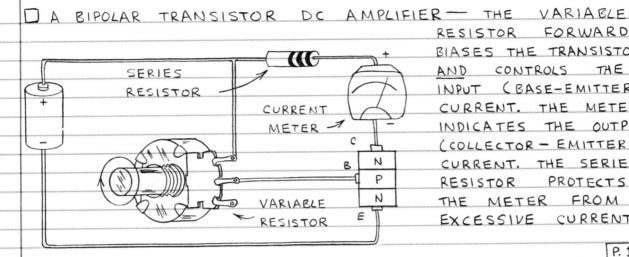

P. 104 →

IN A WORKING CIRCUIT, THE VARIABLE RESISTOR MAY BE IN SERIES WITH A SECOND COMPONENT HAVING A RESISTANCE THAT VARIES WITH TEMPERATURE, LIGHT, MOISTURE, ETC. (WATER IS THE VARIABLE RESISTANCE IN THE MOISTURE METER ON P. 104.) WHEN THE INPUT SIGNAL CHANGES RAPIDLY, AN AC AMPLIFIER SUCH AS THE ONE BELOW IS USED.

☐ **A BIPOLAR TRANSISTOR AC AMPLIFIER** — THIS IS THE SIMPLEST OF SEVERAL BASIC AC AMPLIFIERS. THE INPUT CAPACITOR BLOCKS ANY DC IN THE INPUT SIGNAL.

THE LOAD RESISTOR CAUSES THE OUTPUT CURRENT TO BECOME A VOLTAGE ($V = I \times R$).

P. 122 →

THE OUTPUT SIGNAL IS INVERTED.

THE BIAS RESISTOR IS SELECTED TO GIVE AN OUTPUT VOLTAGE OF ABOUT HALF THE BATTERY VOLTAGE. THE AMPLIFIED SIGNAL "RIDES" ON THIS STEADY OUTPUT VOLTAGE AND VARIES ABOVE AND BELOW IT. (WITHOUT THE BIAS RESISTOR, ONLY THE POSITIVE HALF OF THE INPUT SIGNAL ABOVE 0.6 VOLT (SEE P. 45) WILL BE AMPLIFIED. THIS WILL CAUSE SEVERE DISTORTION.) TO SEE ONE WAY A WORKING VERSION OF THIS AMPLIFIER IS USED, TURN NOW TO P. 122 AND LOOK AT THE OUTPUT SECTION OF THE LIGHTWAVE TRANSMITTER.

FIELD-EFFECT TRANSISTORS

FIELD-EFFECT TRANSISTORS (OR FETs) HAVE BECOME MORE IMPORTANT THAN BIPOLAR TRANSISTORS. THEY ARE EASY TO MAKE AND REQUIRE LESS SILICON. THERE ARE TWO MAJOR FET FAMILIES, JUNCTION AND METAL-OXIDE-SEMICONDUCTOR. IN BOTH KINDS AN OUTPUT CURRENT IS CONTROLLED BY A SMALL INPUT VOLTAGE AND PRACTICALLY NO INPUT CURRENT!

JUNCTION FETs

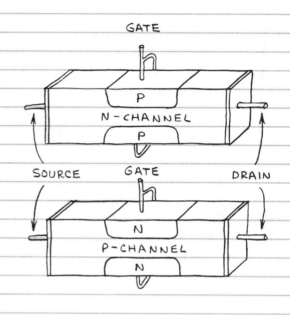

THE TWO MAIN KINDS OF FETs ARE N-CHANNEL AND P-CHANNEL. THE CHANNEL IS LIKE A SILICON RESISTOR THAT CONDUCTS CURRENT MOVING FROM THE SOURCE TO THE DRAIN. A VOLTAGE AT THE GATE INCREASES THE CHANNEL RESISTANCE AND REDUCES THE DRAIN-SOURCE CURRENT. THEREFORE THE FET CAN BE USED AS AN AMPLIFIER OR A SWITCH.

☐ JUNCTION FET OPERATION— THE ARRANGEMENT BELOW SHOWS HOW AN N-CHANNEL FET WORKS. A NEGATIVE GATE VOLTAGE CREATES TWO HIGH RESISTANCE REGIONS (THE FIELD) IN THE CHANNEL ADJACENT TO THE P-TYPE SILICON. MORE GATE VOLTAGE WILL CAUSE THE FIELDS TO MERGE TOGETHER AND COMPLETELY BLOCK THE CURRENT. THE GATE-CHANNEL RESISTANCE IS VERY HIGH.

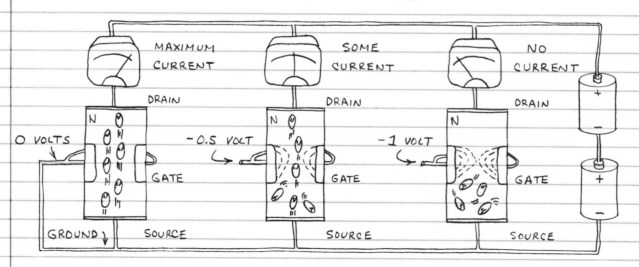

☐ **MORE ABOUT JUNCTION FETs** — SINCE THEY ARE VOLTAGE CONTROLLED, JUNCTION FETs (OR JFETs) HAVE IMPORTANT ADVANTAGES OVER CURRENT-CONTROLLED BIPOLAR TRANSISTORS:

1. THE GATE-CHANNEL RESISTANCE OF A JFET IS VERY HIGH (MILLIONS OF OHMS). THEREFORE THE JFET HAS LITTLE OR NO EFFECT ON EXTERNAL COMPONENTS OR CIRCUITS CONNECTED TO ITS GATE.

2. THE VERY HIGH GATE-CHANNEL RESISTANCE MEANS PRACTICALLY NO CURRENT FLOWS IN THE GATE CIRCUIT. (WHY IS THE RESISTANCE SO HIGH? THE GATE AND CHANNEL FORM A DIODE. SO LONG AS THE INPUT SIGNAL <u>REVERSE</u> BIASES THIS DIODE, THE GATE HAS VERY HIGH INPUT RESISTANCE.)

3. LIKE BIPOLAR TRANSISTORS, JFETs CAN BE DAMAGED OR DESTROYED BY EXCESSIVE CURRENT OR VOLTAGE.

☐ **KINDS OF JUNCTION FETs** — JFETs ARE USED IN MANY DIFFERENT APPLICATIONS. SINCE THEY CANNOT BE USED FOR HIGH POWER ROLES, MOST ARE INSTALLED IN SMALL PLASTIC OR METAL PACKAGES. HERE ARE THE MAIN CATEGORIES:

SMALL SIGNAL AND SWITCHING.

SMALL SIGNAL JFETS ARE USED AT THE INPUT STAGE OF AMPLIFIERS TO PROVIDE A HIGH RESISTANCE INPUT. THEY ARE ALSO USED AS SWITCHES.

HIGH FREQUENCY.

HIGH FREQUENCY JFETS ARE USED TO AMPLIFY OR PRODUCE HIGH FREQUENCY SIGNALS.

☐ **JUNCTION FET SYMBOLS** — GATES INTERNALLY CONNECTED.

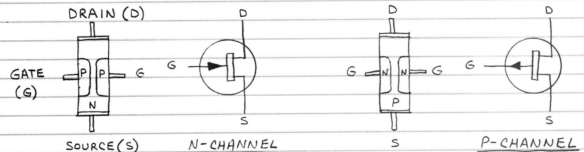

METAL-OXIDE-SEMICONDUCTOR FETs

THE METAL-OXIDE-SEMICONDUCTOR FET (OR MOSFET) HAS BECOME THE MOST IMPORTANT TRANSISTOR. MOST MICROCOMPUTER AND MEMORY INTEGRATED CIRCUITS ARE ARRAYS OF THOUSANDS OF MOSFETs ON A SMALL SLIVER OF SILICON. WHY? MOSFETs ARE EASY TO MAKE, THEY CAN BE VERY SMALL, AND SOME MOSFET CIRCUITS CONSUME NEGLIGIBLE POWER. NEW KINDS OF POWER MOSFETs ARE ALSO VERY USEFUL.

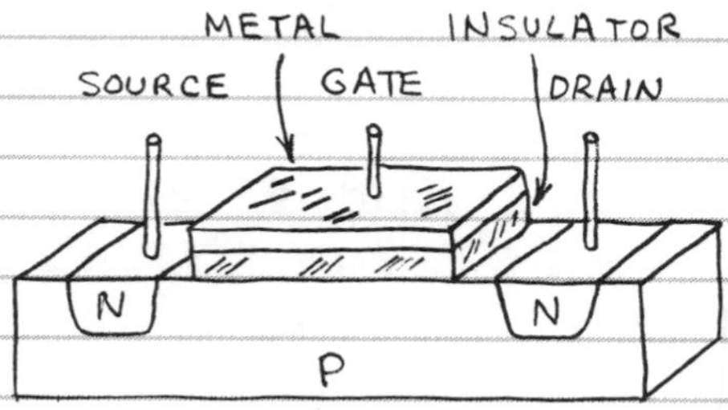

N-MOSFET

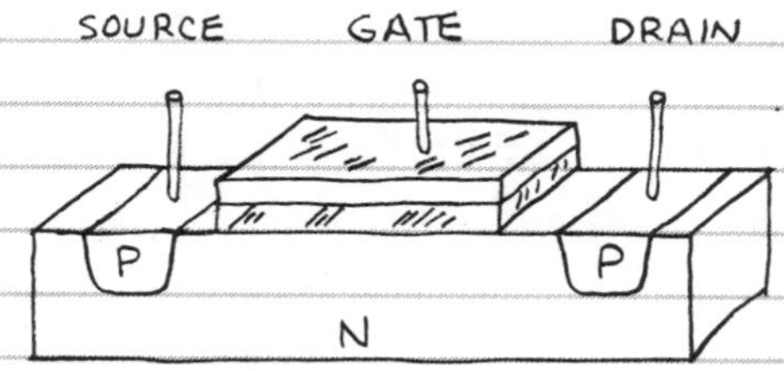

P-MOSFET

□ MOSFET OPERATION — ALL MOSFETs ARE N-TYPE OR P-TYPE. UNLIKE THE JUNCTION FET, THE GATE OF A MOSFET HAS NO ELECTRICAL CONTACT WITH THE SOURCE AND DRAIN. A GLASS-LIKE LAYER OF SILICON-DIOXIDE (AN INSULATOR) SEPARATES THE GATE'S METAL CONTACT FROM THE REST OF THE TRANSISTOR.

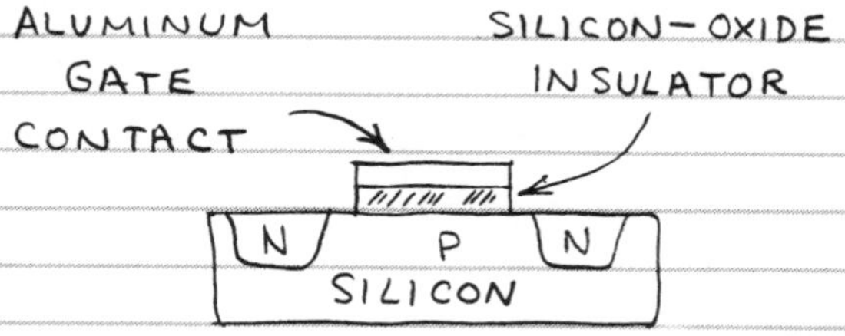

A POSITIVE GATE VOLTAGE ATTRACTS ELECTRONS TO THE REGION BELOW THE GATE. THIS CREATES A THIN N-TYPE CHANNEL IN THE P-TYPE SILICON BETWEEN THE SOURCE AND DRAIN. CURRENT CAN THEN FLOW THROUGH THE CHANNEL. THE GATE VOLTAGE DETERMINES THE RESISTANCE OF THE CHANNEL.

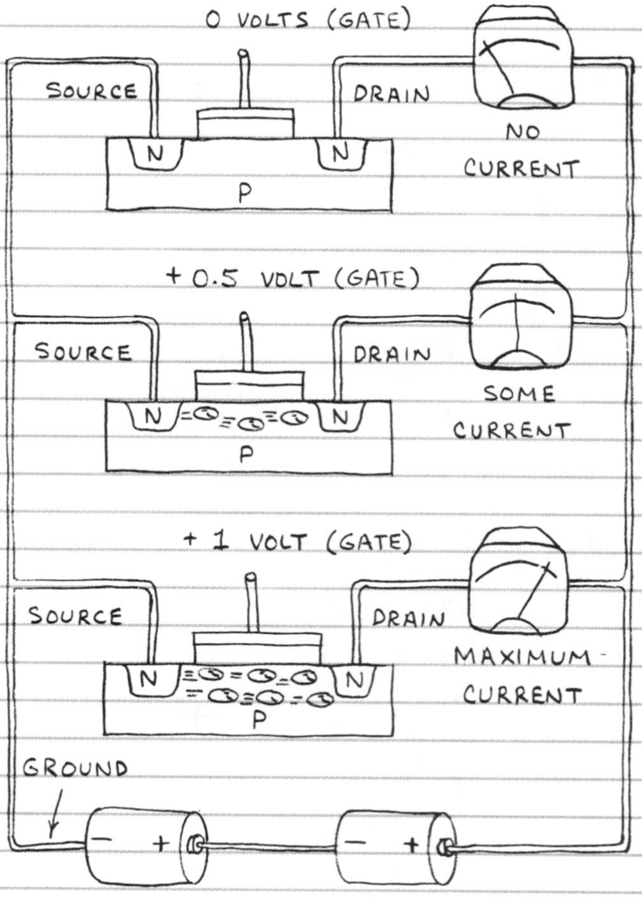

☐ MORE ABOUT MOSFETs — THE INPUT RESISTANCE OF THE MOSFET IS THE HIGHEST OF ANY TRANSISTOR. THIS AND OTHER FACTORS GIVE MOSFETs IMPORTANT ADVANTAGES:

1. THE GATE-CHANNEL RESISTANCE IS ALMOST INFINITE (TYPICALLY 1,000,000,000,000,000-OHMS). THIS MEANS THE GATE PULLS NO CURRENT FROM EXTERNAL CIRCUITS. (WELL, IT MAY BORROW A FEW TRILLIONTHS OF AN AMPERE.)

2. MOSFETs CAN FUNCTION AS VOLTAGE-CONTROLLED VARIABLE RESISTORS. THE GATE VOLTAGE CONTROLS CHANNEL RESISTANCE.

3. NEW KINDS OF MOSFETs CAN SWITCH VERY HIGH CURRENTS IN A FEW BILLIONTHS OF A SECOND.

☐ CAUTION — BECAUSE THE GLASS-LIKE SILICON OXIDE LAYER BELOW THE GATE IS SO THIN, IT CAN BE PIERCED BY TOO MUCH VOLTAGE OR EVEN STATIC ELECTRICITY. EVEN THE STATIC CHARGE GENERATED BY CLOTHING OR A CELLOPHANE WRAPPER CAN ZAP THE GATE OF A MOSFET!

ZAPPED MOSFET   CAUTION SYMBOL

☐ KINDS OF MOSFETs — LIKE JFETs, MOSFETs INSTALLED IN SMALL METAL OR PLASTIC PACKAGES ARE USED TO GIVE AMPLIFIERS AN ULTRA-HIGH INPUT RESISTANCE. THEY ARE ALSO USED AS VOLTAGE CONTROLLED RESISTORS AND SWITCHES. THE MOST IMPORTANT CATEGORY HAS BECOME:

POWER.

POWER MOSFETs ALLOW A FEW VOLTS TO SWITCH OR AMPLIFY MANY AMPERES AT VERY FAST SPEEDS.

☐ MOSFET SYMBOLS — THESE ARE THE MOST COMMON.

N-MOSFET   P-MOSFET

HOW FETs ARE USED

FIELD-EFFECT TRANSISTORS ARE USED AS AMPLIFIERS, SWITCHES AND VOLTAGE-CONTROLLED RESISTORS. HERE ARE SOME TYPICAL CIRCUIT ARRANGEMENTS.

☐ A JFET ELECTROMETER — THIS ULTRA-SIMPLE CIRCUIT IS THE ELECTRONIC VERSION OF THE ELECTROSCOPE. THE GATE LEAD OF AN N-CHANNEL JFET IS LEFT DISCONNECTED. NORMALLY A CURRENT FLOWS FROM SOURCE TO DRAIN. WHEN A <u>NEGATIVELY</u> CHARGED OBJECT (LIKE A PLASTIC COMB THAT'S BEEN STROKED THROUGH YOUR HAIR) IS PLACED NEAR THE GATE, THE CURRENT FLOW IS REDUCED OR STOPPED.

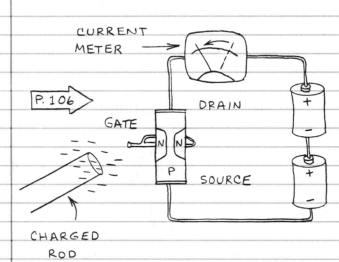

☐ A MOSFET LAMP DRIVER — THIS CIRCUIT SHOWS HOW A POWER MOSFET CAN BE USED TO SWITCH ON A LAMP OR OTHER DC POWERED DEVICE. SINCE THE POWER MOSFET HAS AN ALMOST INFINITE INPUT RESISTANCE, THE SWITCH CAN BE REPLACED BY A <u>TINY</u> INPUT SIGNAL.

☐ A MOSFET LAMP DIMMER — THIS CIRCUIT USES A POWER MOSFET AS A VOLTAGE CONTROLLED RESISTOR.

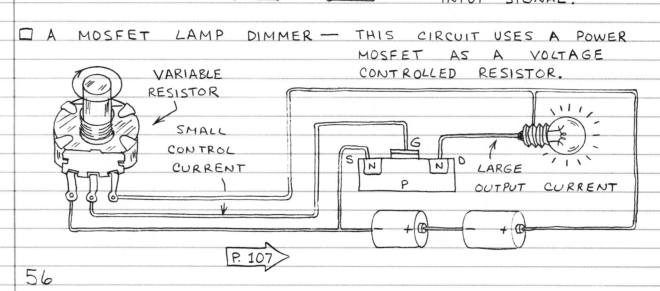

THE UNIJUNCTION TRANSISTOR

The unijunction transistor (UJT) is not a true transistor. It's more like a diode with two cathode connections. It works like a voltage-controlled switch and does not amplify.

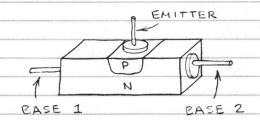

☐ **UJT OPERATION** —
Normally a small current will flow from base 1 to base 2. When the voltage applied to the emitter reaches a certain threshold (several volts), the UJT switches on and a high current flows from base 1 to the emitter. Below the threshold voltage, no current flows from base 1 to the emitter.

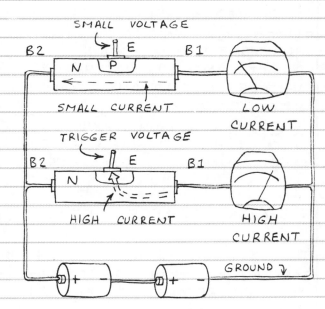

☐ **UJT SYMBOL** —
The symbol for the UJT resembles that of a JFET.

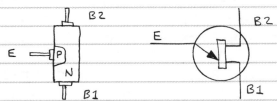

HOW UNIJUNCTION TRANSISTORS ARE USED

This arrangement allows a UJT to flash a light-emitting diode (LED). Current flows into the capacitor until the UJT's trigger voltage is reached. The current in the capacitor is then "dumped" through the LED. The LED glows until the capacitor is discharged. The charge-discharge cycle then repeats.

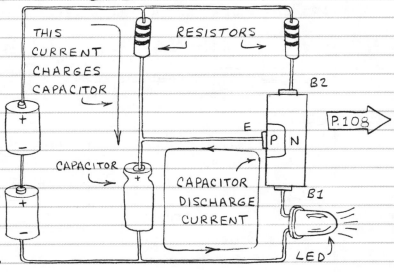

57

THE THYRISTOR

Thyristors are semiconductor devices with three leads. A small current at one lead will allow a much larger current to flow through the other two leads. The controlled current is either ON or OFF. Therefore thyristors do not amplify fluctuating signals like transistors do. Instead they are <u>solid-state switches</u>. There are two families of thyristors, silicon-controlled rectifiers (SCRs) and TRIACs. SCRs switch direct current and TRIACs switch alternating current.

SILICON-CONTROLLED RECTIFIERS (SCRs)

The SCR is similar to a bipolar transistor with a fourth layer and therefore <u>three</u> PN junctions. It is sometimes called a 4-layer PNPN diode since it passes a current in only one direction.

☐ SCR OPERATION — If the anode of an SCR is made more positive than the cathode the two outermost PN junctions are forward biased. The middle PN junction, however, is reverse biased and current cannot flow. A small gate current forward biases the middle PN junction and allows a much larger current to flow through the device. The SCR stays <u>ON</u> even if the gate current is removed! (until power is disconnected.)

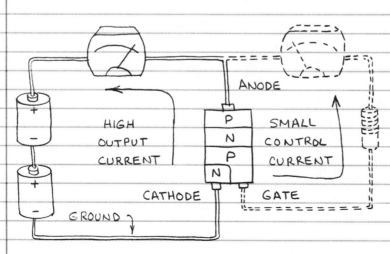

☐ KINDS OF SCRs — SCRs ARE CATEGORIZED ACCORDING TO THE CURRENT THEY CAN SWITCH. HERE ARE THREE GENERAL CATEGORIES (MANY OTHER CASE STYLES ARE AVAILABLE):

LOW CURRENT.

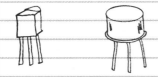

LOW CURRENT SCRs INCLUDE THOSE THAT SWITCH UP TO 1-AMPERE AT UP TO 100-VOLTS.

MEDIUM CURRENT.

THESE SCRs SWITCH UP TO 10-AMPERES AT UP TO SEVERAL HUNDRED VOLTS. ONE COMMON USE IS SOLID-STATE SWITCHING FOR AUTO ENGINES.

HIGH CURRENT.

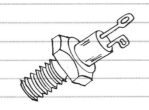

THESE SCRs CAN SWITCH UP TO 2,500-AMPERES AT UP TO SEVERAL THOUSAND VOLTS! THEY CONTROL MOTORS, LIGHTS, APPLIANCES, ETC.

☐ SCR SYMBOL.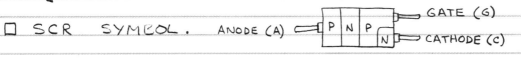

HOW SCRs ARE USED

THIS ARRANGEMENT SHOWS HOW AN SCR IS USED TO SWITCH ON AN INCANDESCENT LAMP. OTHER DEVICES CAN ALSO BE CONTROLLED.

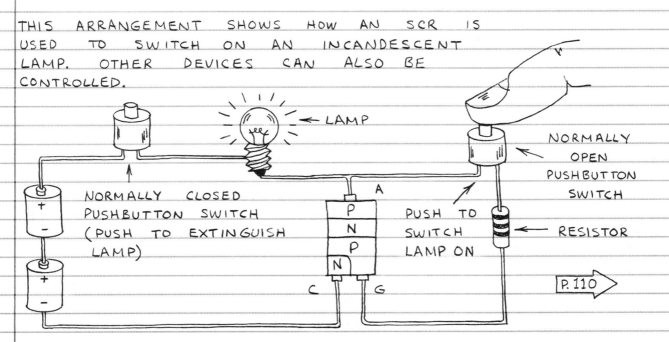

P. 110 →

59

TRIACS

THE TRIAC IS EQUIVALENT TO TWO SCRs CONNECTED IN PARALLEL. THIS MEANS TRIACS CAN SWITCH BOTH DIRECT AND ALTERNATING CURRENT. NOTICE THAT THE TRIAC HAS FIVE LAYERS PLUS AN EXTRA N-TYPE REGION. ALSO NOTE HOW ALL THREE LEADS MAKE CONTACT WITH TWO LAYERS.

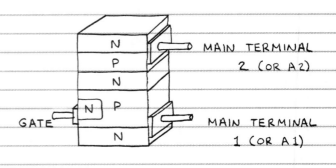

☐ TRIAC OPERATION — THE TWO PARALLEL SCRs THAT FORM A TRIAC FACE IN OPPOSITE DIRECTIONS (REVERSE-PARALLEL). WHEN USED TO SWITCH ALTERNATING CURRENT, THE TRIAC STAYS ON ONLY WHEN THE GATE RECEIVES CURRENT. REMOVE THE GATE CURRENT AND IT SWITCHES OFF WHEN THE AC PASSES THROUGH 0 VOLTS.

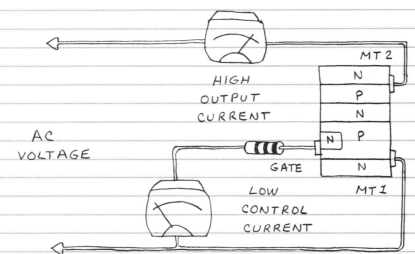

☐ KINDS OF TRIACS — TRIACS, LIKE SCRs, ARE CATEGORIZED ACCORDING TO THE CURRENT THEY CAN SWITCH. TRIACS DON'T HAVE THE VERY HIGH POWER CAPABILITY OF HIGH CURRENT SCRs. HERE ARE TWO CATEGORIES:

LOW CURRENT.

LOW CURRENT TRIACS SWITCH UP TO 1-AMPERE AT UP TO SEVERAL HUNDRED VOLTS. OTHER CASE STYLES ALSO USED.

MEDIUM CURRENT.

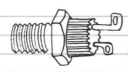

THESE TRIACS SWITCH UP TO 40-AMPERES AT UP TO 1,000-VOLTS. MANY CASE STYLES ARE AVAILABLE.

☐ TRIAC SYMBOL — REMEMBER, THE TRIAC IS THE SAME AS TWO REVERSE-PARALLEL SCRs:

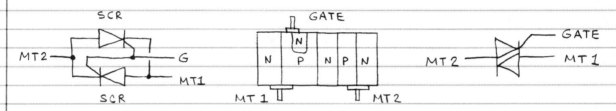

HOW TRIACS ARE USED

THIS ARRANGEMENT SHOWS HOW A TRIAC CAN SWITCH ON A LAMP POWERED BY HOUSEHOLD LINE CURRENT. MOTORS AND OTHER DEVICES CAN ALSO BE CONTROLLED.

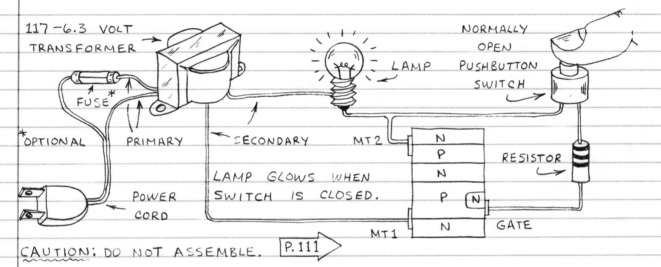

CAUTION: DO NOT ASSEMBLE. P.111 →

TWO-LEAD THYRISTORS

AN SCR OR TRIAC WILL SWITCH ON <u>WITHOUT A GATE</u> SIGNAL IF THE VOLTAGE ACROSS ITS OTHER TWO LEADS REACHES A CERTAIN LEVEL (THE <u>BREAKDOWN VOLTAGE</u>). THIS SELF-SWITCHING ABILITY MAKES POSSIBLE TWO-LEAD THYRISTORS.

FOUR-LAYER DIODE.

A FOUR-LAYER DIODE IS AN SCR WITHOUT A GATE. IT SWITCHES DC VOLTAGE.

DIAC.

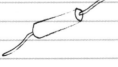

A DIAC IS A THREE-LAYER DEVICE SIMILAR TO A PNP JUNCTION TRANSISTOR WITHOUT A BASE LEAD. IT SWITCHES AC VOLTAGE.

4. PHOTONIC SEMICONDUCTORS

PHOTONICS IS THE FAST GROWING FIELD OF ELECTRONICS INVOLVING SEMICONDUCTOR DEVICES THAT EMIT AND DETECT LIGHT. BEFORE LOOKING AT SOME PHOTONIC COMPONENTS, LET'S TAKE A QUICK LOOK AT SOME FACTS ABOUT LIGHT.

LIGHT

"LET THERE BE LIGHT..."

LIGHT IS COMPOSED OF PARTICLES CALLED <u>PHOTONS</u> THAT BEHAVE LIKE WAVES OF ENERGY. PHOTONS ARE <u>NOT</u> NECESSARILY VISIBLE AND ONLY THOSE YOU CAN SEE ARE COLLECTIVELY CALLED LIGHT. PHOTONS ARE PRODUCED WHEN AN ELECTRON THAT'S BEEN EXCITED TO A HIGHER THAN NORMAL ENERGY LEVEL FALLS BACK TO ITS NORMAL LEVEL.

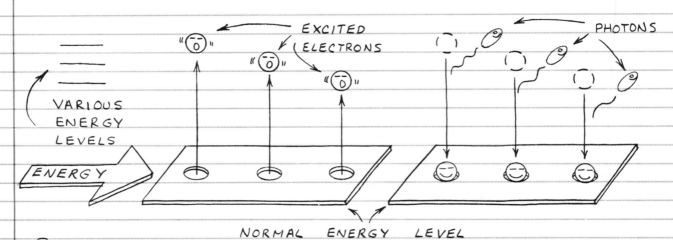

REMEMBER, PHOTONS ACT LIKE WAVES. THE DISTANCE BETWEEN CRESTS IS THE <u>WAVELENGTH</u>. ELECTRONS EXCITED TO HIGHER ENERGY LEVELS EMIT PHOTONS WITH SHORTER WAVELENGTHS THAN ELECTRONS EXCITED TO LOWER LEVELS.

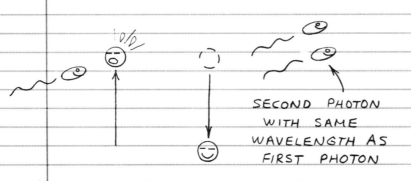

EXCITED ELECTRONS CAN RESUME THEIR NORMAL LEVEL <u>SPONTANEOUSLY</u>. OR A PHOTON WITH THE PROPER WAVELENGTH CAN <u>STIMULATE</u> AN EXCITED ELECTRON TO RETURN TO ITS NORMAL LEVEL.

☐ THE ELECTROMAGNETIC SPECTRUM — VISIBLE LIGHT IS A FORM OF ELECTROMAGNETIC RADIATION. THE WAVELENGTH OF LIGHT IS SPECIFIED IN NANOMETERS (1-NANOMETER IS A BILLIONTH OF A METER). THE DIAGRAM BELOW SHOWS THE RELATIONSHIP OF LIGHT TO OTHER FORMS OF ELECTROMAGNETIC RADIATION.

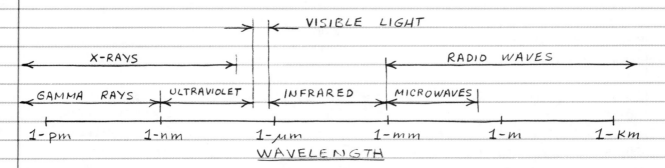

1-pm = 1 PICOMETER   (0.000 000 000 001 METER)
1-nm = 1 NANOMETER   (0.000 000 001 METER)
1-µm = 1 MICROMETER   (0.000 001 METER)
1-mm = 1 MILLIMETER   (0.001 METER)
1-m = 1 METER   (39.37 INCHES)
1-km = 1 KILOMETER   (1000 METERS)

THESE LINES ARE ONE MILLIMETER (1-mm) APART.

☐ THE OPTICAL SPECTRUM — ULTRAVIOLET, VISIBLE AND INFRARED RADIATION ARE TOGETHER CALLED THE OPTICAL SPECTRUM. HERE'S AN EXPANDED DIAGRAM OF THE OPTICAL SPECTRUM:

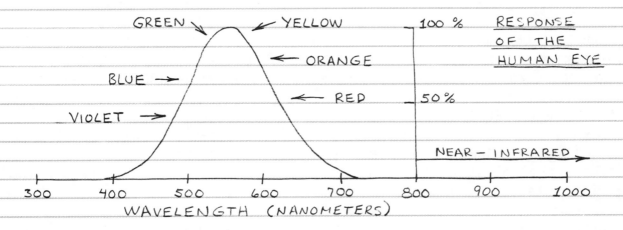

MANY PHOTONIC SEMICONDUCTORS EMIT OR DETECT NEAR-INFRARED RADIATION. SILICON, FOR EXAMPLE, CAN DETECT VISIBLE LIGHT. BUT IT IS MOST SENSITIVE TO NEAR-INFRARED AT ABOUT 880 nm. BECAUSE SO MANY PHOTONIC COMPONENTS CAN OPERATE IN BOTH THE VISIBLE AND NEAR-INFRARED, IT'S COMMON TO REFER TO NEAR-INFRARED AS LIGHT.

OPTICAL COMPONENTS

OPTICAL COMPONENTS CONDUCT, BEND OR CHANGE THE CHARACTERISTICS OF LIGHT. SEVERAL ARE VERY IMPORTANT IN MANY APPLICATIONS OF PHOTONIC SEMICONDUCTORS:

1. **FILTERS** TRANSMIT ONLY A NARROW BAND OF OPTICAL WAVELENGTHS.

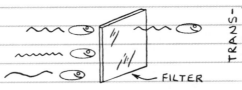

2. **REFLECTORS** REFLECT SOME OR MOST OF AN ONCOMING LIGHT BEAM. SOME LIGHT MAY OR MAY NOT BE TRANSMITTED. THOSE WITH A VERY SMOOTH SURFACE (LIKE MIRRORS) ARE CALLED **SPECULAR REFLECTORS**.

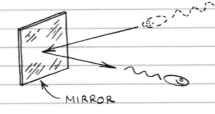

3. **BEAMSPLITTERS** REFLECT PART OF AN ONCOMING LIGHT BEAM AND TRANSMIT THE REMAINDER. A GLASS MICROSCOPE SLIDE MAKES A GOOD BEAM SPLITTER. (EACH SURFACE REFLECTS 4%.)

4. **LENSES** BEND LIGHT. THE MOST IMPORTANT ARE:

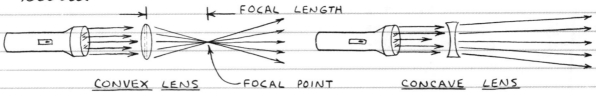

CONVEX LENSES ARE <u>OFTEN</u> USED IN CONJUNCTION WITH SEMICONDUCTOR LIGHT SOURCES AND DETECTORS. FOR EXAMPLE, THEY CAN COLLECT AND FOCUS LIGHT ONTO A MINIATURE DETECTOR.

5. **OPTICAL FIBERS** ARE THIN, FLEXIBLE STRANDS OF HIGHLY TRANSPARENT GLASS OR PLASTIC THAT CONDUCT LIGHT. THE LIGHT TRAVELS THROUGH A **CORE** SURROUNDED BY A THIN **CLADDING**. PLASTIC FIBERS ARE INEXPENSIVE. GLASS FIBERS ARE MUCH MORE TRANSPARENT. BOTH KINDS TRANSMIT SOME WAVELENGTHS MUCH BETTER THAN OTHERS. HIGH QUALITY FIBERS ARE USED TO SEND TELEPHONE CALLS AND COMPUTER DATA VIA PULSES OF LIGHT.

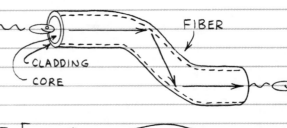

GLASS IS HUNDREDS OF TIMES CLEARER.

HOW CONVEX LENSES ARE USED

MANY SEMICONDUCTOR LIGHT SOURCES AND DETECTORS ARE EQUIPPED WITH A **BUILT-IN** CONVEX LENS. THIS PAGE EXPLAINS WHY AND SHOWS HOW EXTERNAL LENS ARE USED WITH SOURCES AND DETECTORS.

☐ THE INVERSE SQUARE LAW.

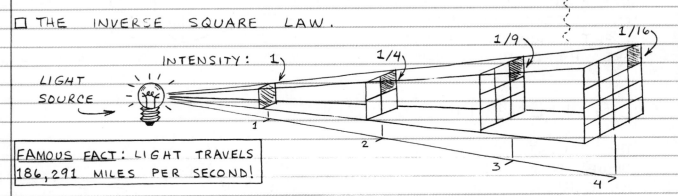

FAMOUS FACT: LIGHT TRAVELS 186,291 MILES PER SECOND!

AS LIGHT FROM A VERY SMALL SOURCE SPREADS OUTWARD, ITS INTENSITY IS INVERSELY PROPORTIONAL TO THE SQUARE OF THE DISTANCE. IN OTHER WORDS, IF THE DISTANCE IS 3, THEN THE INTENSITY IS 1/9 THE INTENSITY WHEN THE DISTANCE IS 1. A CONVEX LENS CAN CANCEL THIS INTENSITY REDUCTION.

☐ THE CONVEX LENS.

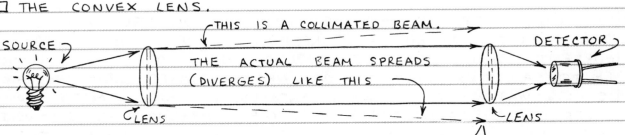

THE BEAM SPREAD ANGLE (THE DIVERGENCE) IN RADIANS* IS THE DIAMETER OF THE SOURCE DIVIDED BY THE FOCAL LENGTH OF THE LENS. THIS MEANS LENSES WITH LONGER FOCAL LENGTHS GIVE NARROWER BEAMS. (BUT LENSES WITH LONG FOCAL LENGTHS COLLECT LESS LIGHT THAN THOSE WITH SHORT FOCAL LENGTHS...)

*ONE RADIAN IS 57.3 DEGREES. (THERE ARE 360° IN A CIRCLE.)

WITH CAREFUL LENS PLACEMENT ALL THE LIGHT WITHIN THE DASHED CIRCLE CAN BE FOCUSED ONTO THE LIGHT SENSITIVE PORTION OF THE DETECTOR. (SURE BEATS THE INVERSE SQUARE LAW!)

SEMICONDUCTOR LIGHT SOURCES

WHEN BOMBARDED BY LIGHT, HEAT, ELECTRONS AND OTHER FORMS OF ENERGY, MOST SEMICONDUCTOR CRYSTALS WILL EMIT VISIBLE OR INFRARED LIGHT. THE BEST SEMICONDUCTOR LIGHT SOURCES, HOWEVER, ARE PN JUNCTION DIODES.

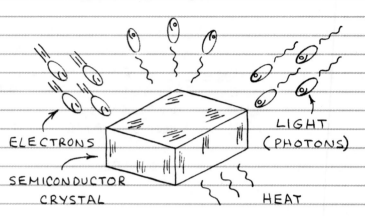

LIGHT EMITTING DIODES

THE LIGHT EMITTING DIODE CONVERTS AN ELECTRICAL CURRENT DIRECTLY INTO LIGHT. THEREFORE THE LIGHT EMITTING DIODE (LED) IS MORE EFFICIENT THAN MANY OTHER LIGHT SOURCES.

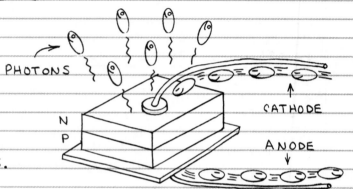

☐ LED OPERATION — THE FORWARD VOLTAGE ACROSS A DIODE MUST EXCEED A THRESHOLD LEVEL BEFORE A CURRENT CAN CROSS THE JUNCTION. FOR SILICON, WHICH EMITS A TINY AMOUNT OF NEAR-INFRARED, THE THRESHOLD IS 0.6-VOLT. FOR GALLIUM ARSENIDE, WHICH EMITS CONSIDERABLE NEAR-INFRARED, THE THRESHOLD IS 1.3-VOLTS. THIS VOLTAGE EXCITES THE ELECTRONS. WHEN THE ELECTRONS CROSS THE JUNCTION AND COMBINE WITH HOLES, THEY EMIT PHOTONS.

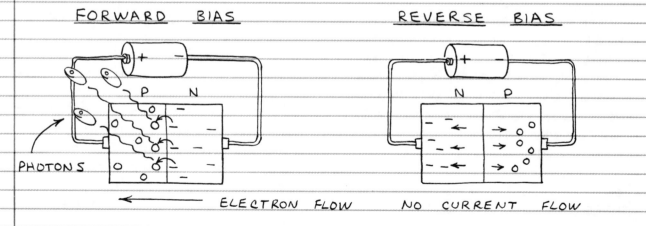

☐ MORE ABOUT LED OPERATION — HERE ARE SOME KEY ASPECTS OF LED OPERATION YOU SHOULD KNOW ABOUT:

1. THE LIGHT EMITTED BY AN INCANDESCENT LAMP CONTAINS MANY WAVELENGTHS. THE LIGHT EMITTED BY AN LED HAS A <u>NARROW</u> WAVELENGTH RANGE. (THIS IS BECAUSE THE ELECTRONS IN THE LED ARE ALL EXCITED TO THE SAME LEVEL.)

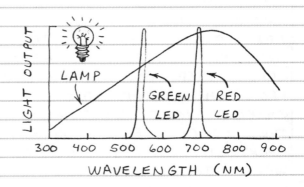

2. WHEN AN LED BEGINS TO CONDUCT, THE <u>VOLTAGE INCREASES GRADUALLY</u> WHILE THE <u>CURRENT INCREASES RAPIDLY</u>. TOO MUCH CURRENT WILL OVERHEAT THE LED AND POSSIBLY SEPARATE THE LEADS OR MELT THE SEMICONDUCTOR CHIP.

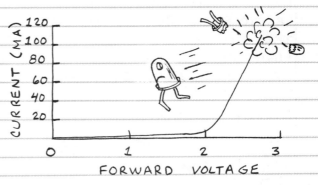

3. THE LIGHT EMITTED BY AN LED IS <u>DIRECTLY PROPORTIONAL</u> TO CURRENT THROUGH THE LED. THIS MEANS LEDS ARE IDEAL FOR TRANSMITTING INFORMATION. THE LIGHT OUTPUT FROM AN OVERHEATED LED WILL SOON <u>DECREASE</u>. THE LED MAY EVEN BE DAMAGED.

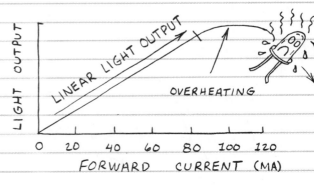

4. THE FORWARD VOLTAGE AND WAVELENGTH OF AN LED ARE DIRECTLY RELATED. THEREFORE IT'S NOT ALWAYS POSSIBLE TO SUBSTITUTE DIFFERENT LEDS WITHOUT CHANGING THE VOLTAGE AND CURRENT. <u>MANY</u> DIFFERENT SEMICONDUCTORS ARE USED TO MAKE VARIOUS LEDS. VISIBLE LIGHT EMITTING LEDS

WAVELENGTH (NM)	VOLTAGE
565 (GREEN)	2.2 – 3.0
590 (YELLOW)	2.2 – 3.0
615 (ORANGE)	1.8 – 2.7
640 (RED)	1.6 – 2.0
690 (RED)	2.2 – 3.0
880 (INFRARED)	2.0 – 2.5
900 (INFRARED)	1.2 – 1.6
940 (INFRARED)	1.3 – 1.7

EMIT UP TO A MILLIWATT OR SO OF POWER. SOME INFRARED LEDS (LIKE 880 NM UNITS) EMIT 15 OR MORE MILLIWATTS! (A FLASHLIGHT EMITS 10 OR MORE MILLIWATTS.)

☐ KINDS OF LEDs — SINCE THE LED IS A LIGHT SOURCE, IT'S HELPFUL TO KNOW WHAT'S INSIDE THE PLASTIC OR METAL LED CASE. SHOWN HERE IS A TYPICAL LED. THE HEAVY LEADS HELP CONDUCT HEAT AWAY FROM THE CHIP. THE REFLECTOR COLLECTS LIGHT EMITTED FROM THE EDGES OF THE CHIP. THE EPOXY IS USUALLY COLORED WHEN THE LED IS A VISIBLE LIGHT EMITTER. LIGHT SCATTERING PARTICLES ARE OFTEN ADDED TO THE EPOXY. THIS DIFFUSES THE LIGHT AND CAUSES THE END OF THE LED TO APPEAR BRIGHTER.

VISIBLE LIGHT LEDs.

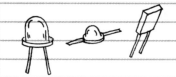

THESE INEXPENSIVE LEDs ARE USED AS INDICATOR LIGHTS. CERTAIN RED LEDs ARE USED TO TRANSMIT INFORMATION. MOST ARE ENCAPSULATED IN EPOXY.

LED DISPLAYS.

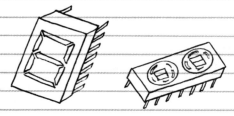

MANY KINDS OF LED READOUTS CAPABLE OF DISPLAYING DIGITS AND CHARACTERS ARE AVAILABLE. THEY ARE MORE RUGGED THAN LIQUID CRYSTAL DISPLAYS, BUT THEY USE MUCH MORE CURRENT.

INFRARED LEDs.

INFRARED LEDs SHOULD BE CALLED INFRARED EMITTING DIODES. THEY ARE USED TO TRANSMIT INFORMATION. THEY ARE ALSO USED IN INTRUSION ALARMS, REMOTE CONTROL DEVICES, ETC. A SPECIAL KIND OF INFRARED LED IS THE DIODE LASER. SOME EMIT SEVERAL WATTS!

☐ LED SYMBOL. BOTH SYMBOLS SHOWN HERE ARE USED.

HOW LEDs ARE USED

LEDs can be powered by continuous current or by brief pulses of current. When operated continuously, the current can be varied to change the light output.

☐ **LED DRIVE CIRCUIT** — Because LEDs are current dependent, it's usually necessary to protect them from excessive current with a <u>series resistor</u>. Some LEDs include a built-in series resistor. <u>MOST DO NOT.</u> It's important to know how to determine the required series resistance (R_s). The formula is:

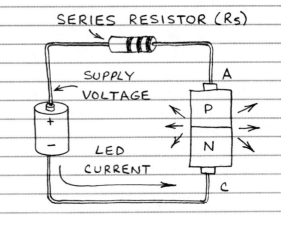

$$R_s = \frac{\text{SUPPLY VOLTAGE} - \text{LED VOLTAGE}}{\text{LED CURRENT}} \quad \text{OR} \quad R_s = \frac{V - V_{LED}}{I_{LED}} \quad \boxed{P.112 \Rightarrow}$$

EXAMPLE: Suppose you want to operate a red LED at a forward current (I_{LED}) of 10-milliamperes from a 5-volt supply (V). V_{LED} is 1.7-volts (from data sheet). Therefore R_s is $(5-1.7)/0.01$ or 330-ohms.

☐ **LED POLARITY INDICATOR** — Two reverse-parallel LEDs form a polarity indicator. <u>BOTH</u> LEDs glow if the tested voltage is AC. The series resistor <u>MUST</u> be used! $\boxed{P.112 \Rightarrow}$

☐ **PULSED LED** — When operated continuously, an infrared LED might have a maximum current of 100-milliamperes. When driven by brief pulses of current, the same LED may safely accept huge 10-ampere pulses!

NOTE: A series resistor may <u>NOT</u> be required if the pulses do not exceed the maximum levels specified for the LED.

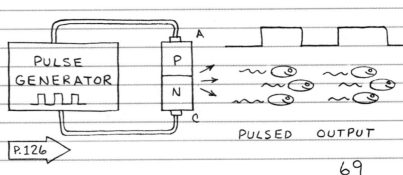

$\boxed{P.126 \Rightarrow}$

SEMICONDUCTOR LIGHT DETECTORS

ENERGY ENTERING A SEMICONDUCTOR CRYSTAL EXCITES ELECTRONS TO HIGHER LEVELS, LEAVING BEHIND HOLES. THESE ELECTRONS AND HOLES CAN RECOMBINE AND EMIT PHOTONS, OR THEY CAN MOVE AWAY FROM ONE ANOTHER AND FORM A CURRENT. THIS IS THE BASIS OF SEMICONDUCTOR LIGHT DETECTORS. THERE ARE TWO MAJOR CLASSES OF SEMICONDUCTOR LIGHT DETECTORS, THOSE WITH AND THOSE WITHOUT PN JUNCTIONS.

PHOTORESISTIVE LIGHT DETECTORS

PHOTORESISTORS ARE SEMICONDUCTOR LIGHT DETECTORS WITHOUT A PN JUNCTION. THEIR RESISTANCE IS VERY HIGH (UP TO MILLIONS OF OHMS) WHEN NO LIGHT IS PRESENT. WHEN ILLUMINATED, THEIR RESISTANCE IS VERY LOW (HUNDREDS OF OHMS).

☐ PHOTORESISTOR OPERATION. THIS PANEL SHOWS HOW A PHOTON CREATES A HOLE-ELECTRON PAIR. AN EXTERNAL VOLTAGE WILL FORCE THE HOLE AND ELECTRON TO MOVE.

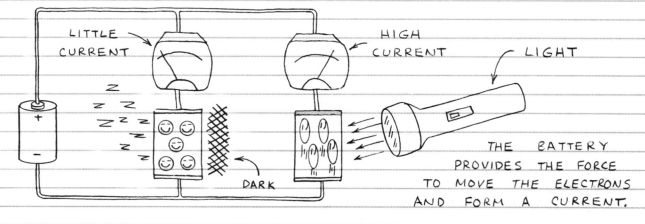

THE BATTERY PROVIDES THE FORCE TO MOVE THE ELECTRONS AND FORM A CURRENT.

☐ MORE ABOUT PHOTORESISTOR OPERATION— HERE ARE SOME IMPORTANT ASPECTS OF PHOTORESISTOR OPERATION:

1. PHOTORESISTORS MAY REQUIRE A FEW MILLISECONDS OR MORE TO FULLY RESPOND TO A CHANGE IN LIGHT INTENSITY (THAT'S PRETTY SLOW). THEY MAY REQUIRE MANY MINUTES TO RETURN TO THEIR NORMAL DARK RESISTANCE WHEN LIGHT IS REMOVED (THE MEMORY EFFECT).

2. THE SEMICONDUCTOR MOST OFTEN USED IN PHOTORESISTORS IS CADMIUM SULFIDE. ITS SENSITIVITY TO LIGHT IS VERY SIMILAR TO THAT OF THE HUMAN EYE! LEAD SULFIDE IS USED TO DETECT INFRA-RED (OUT TO 3-MICROMETERS).

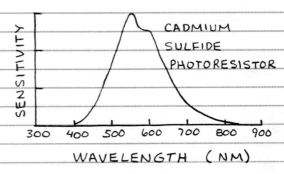

☐ KINDS OF PHOTORESISTORS— MANY DIFFERENT KINDS ARE AVAILABLE. IN MOST THE LIGHT SENSITIVE SEMICONDUCTOR IS COATED BETWEEN INTERLEAVED ELECTRODES TO INCREASE THE EXPOSED SURFACE. A PLASTIC OR GLASS WINDOW MAY OR MAY NOT BE USED.

☐ PHOTORESISTOR SYMBOL. BOTH SYMBOLS SHOWN HERE ARE USED.

 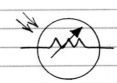

HOW PHOTORESISTORS ARE USED

P. 114 →

PHOTORESISTORS ARE USED IN LIGHT CON-TROLLED RELAYS AND LIGHT METERS.

☐ LIGHT METER. THE ARRANGEMENT SHOWN HERE INDICATES ON A CURRENT METER THE INTENSITY OF LIGHT ILLUMINATING A CADMIUM SULFIDE PHOTORESISTOR.

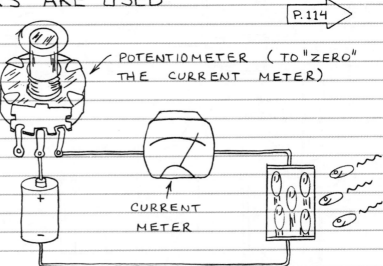

POTENTIOMETER (TO "ZERO" THE CURRENT METER)

CURRENT METER

71

PN JUNCTION LIGHT DETECTORS

PN JUNCTION LIGHT DETECTORS FORM THE LARGEST FAMILY OF PHOTONIC SEMICONDUCTORS. MOST ARE MADE FROM SILICON AND CAN DETECT BOTH VISIBLE LIGHT AND NEAR-INFRARED.

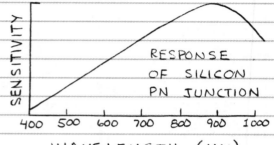

PHOTODIODES

ALL PN JUNCTIONS ARE LIGHT SENSITIVE. PHOTODIODES ARE PN JUNCTIONS SPECIFICALLY DESIGNED FOR LIGHT DETECTION. THEY ARE USED IN CAMERAS, INTRUSION ALARMS, LIGHTWAVE COMMUNICATORS, ETC.

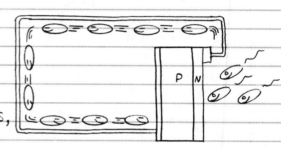

☐ PHOTODIODE OPERATION.
A PHOTON WILL CREATE A HOLE-ELECTRON PAIR AT A PN JUNCTION. A CURRENT WILL FLOW IF THE TWO SIDES OF THE JUNCTION ARE CONNECTED. TWO OPERATING MODES ARE POSSIBLE:

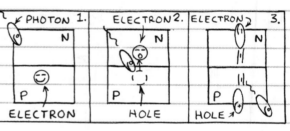

1. PHOTOVOLTAIC OPERATION — HERE THE PHOTODIODE BECOMES A CURRENT SOURCE WHEN IT IS ILLUMINATED.

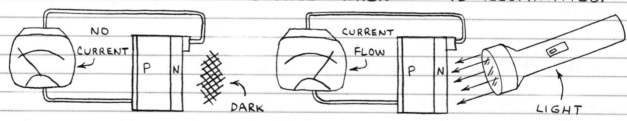

2. PHOTOCONDUCTIVE OPERATION — HERE THE PHOTODIODE IS REVERSE-BIASED. A CURRENT FLOWS WHEN THE PN JUNCTION IS ILLUMINATED. (WHEN DARK, A TINY CURRENT CALLED THE DARK CURRENT WILL FLOW.)

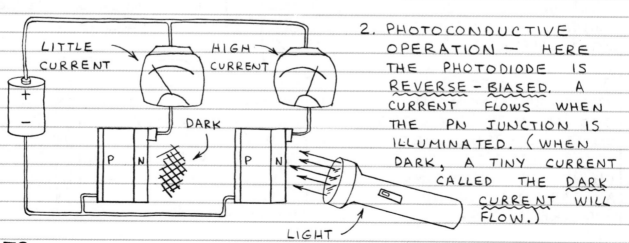

☐ **KINDS OF PHOTODIODES.**
SHOWN HERE IS A TYPICAL
PHOTODIODE. MANY OTHER
CASE STYLES ARE ALSO USED
(PLASTIC HOUSING, BUILT-IN
LENSES AND FILTERS, ETC.).
THE MOST IMPORTANT DIS-
TINCTION IS THE SIZE OF
THE SEMICONDUCTOR CHIP.
SPECIALIZED CHIP DESIGNS
MAY BE USED TO GIVE
BETTER RESPONSE TO CERTAIN
WAVELENGTHS OF LIGHT.

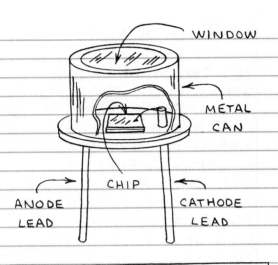

FAMOUS FACT: LEDs CAN BOTH EMIT AND DETECT LIGHT!

SMALL AREA PHOTODIODES.

THESE PHOTODIODES HAVE <u>VERY</u> FAST
RESPONSE TIMES WHEN USED IN THE
REVERSE-BIASED PHOTOCONDUCTIVE MODE.

LARGE AREA PHOTODIODES.

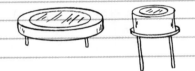

THOUGH SLOWER RESPONDING THAN SMALL
AREA PHOTODIODES, THEIR LARGE AREA
PROVIDES HIGH SENSITIVITY.

☐ **PHOTODIODE SYMBOL.**
BOTH SYMBOLS SHOWN
HERE ARE USED.

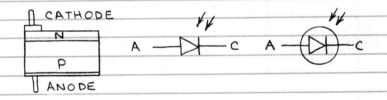

HOW PHOTODIODES ARE USED

PHOTODIODES ARE
COMMONLY USED TO
DETECT FAST PULSES
OF NEAR INFRARED
(AS IN LIGHTWAVE
COMMUNICATIONS).

☐ **LIGHT METER.**
THIS ARRANGEMENT
PROVIDES A BASIC
PHOTOCONDUCTIVE MODE
LIGHT METER. ITS
RESPONSE IS <u>VERY</u> LINEAR.

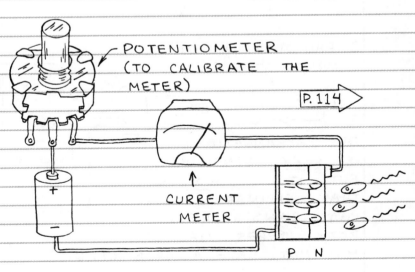

P.114

PHOTOTRANSISTORS

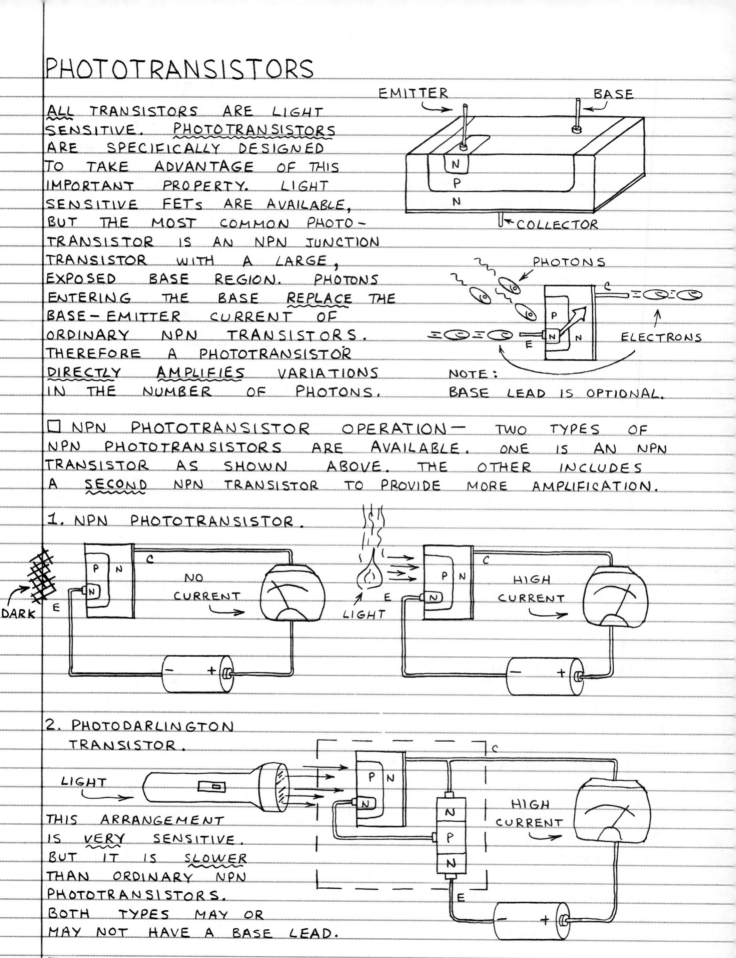

ALL TRANSISTORS ARE LIGHT SENSITIVE. PHOTOTRANSISTORS ARE SPECIFICALLY DESIGNED TO TAKE ADVANTAGE OF THIS IMPORTANT PROPERTY. LIGHT SENSITIVE FETs ARE AVAILABLE, BUT THE MOST COMMON PHOTOTRANSISTOR IS AN NPN JUNCTION TRANSISTOR WITH A LARGE, EXPOSED BASE REGION. PHOTONS ENTERING THE BASE REPLACE THE BASE-EMITTER CURRENT OF ORDINARY NPN TRANSISTORS. THEREFORE A PHOTOTRANSISTOR DIRECTLY AMPLIFIES VARIATIONS IN THE NUMBER OF PHOTONS.

NOTE: BASE LEAD IS OPTIONAL.

☐ NPN PHOTOTRANSISTOR OPERATION— TWO TYPES OF NPN PHOTOTRANSISTORS ARE AVAILABLE. ONE IS AN NPN TRANSISTOR AS SHOWN ABOVE. THE OTHER INCLUDES A SECOND NPN TRANSISTOR TO PROVIDE MORE AMPLIFICATION.

1. NPN PHOTOTRANSISTOR.

2. PHOTODARLINGTON TRANSISTOR.

THIS ARRANGEMENT IS VERY SENSITIVE. BUT IT IS SLOWER THAN ORDINARY NPN PHOTOTRANSISTORS. BOTH TYPES MAY OR MAY NOT HAVE A BASE LEAD.

☐ **KINDS OF PHOTOTRANSISTORS.**
SHOWN HERE IS A TYPICAL
LOW COST NPN PHOTOTRANSISTOR.
MANY OTHER CASE STYLES
ARE ALSO USED (METAL CANS,
GLASS LENSES, FLAT WINDOWS,
ETC.). IMPORTANT: THE BASE
LEAD MAY OR MAY NOT BE
PRESENT. MANY PHOTOTRANSISTOR
CIRCUITS DO NOT USE THE
BASE CONNECTION.

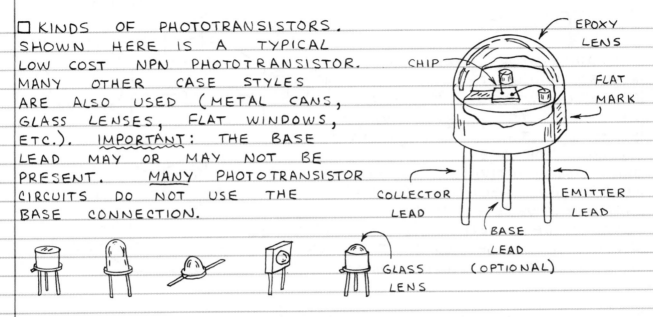

THESE ARE TYPICAL PHOTOTRANSISTORS.

☐ **PHOTOTRANSISTOR SYMBOLS.**

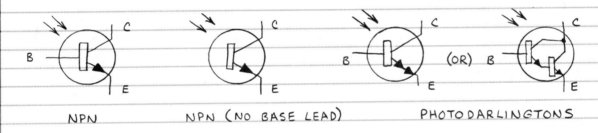

HOW PHOTOTRANSISTORS ARE USED

PHOTOTRANSISTORS ARE OFTEN
USED TO DETECT FLUCTUATING (AC)
LIGHT SIGNALS. THIS ARRANGEMENT
USES A STEADY (DC)
LIGHT TO ENERGIZE
A RELAY.

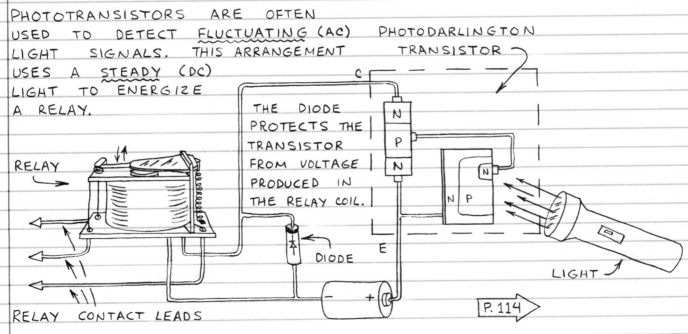

P. 114 →

75

PHOTOTHYRISTORS

PHOTOTHYRISTORS ARE VARIOUS KINDS OF LIGHT-ACTIVATED THYRISTORS. YOU CAN THINK OF THEM AS <u>LIGHT-ACTIVATED SWITCHES</u>. THE MOST IMPORTANT MEMBER OF THE FAMILY IS THE LIGHT-ACTIVATED SILICON CONTROLLED RECTIFIER (LASCR). LIGHT ACTIVATED TRIACS ARE ALSO MADE. NEITHER CAN SWITCH AS MUCH CURRENT AS CONVENTIONAL THYRISTORS.

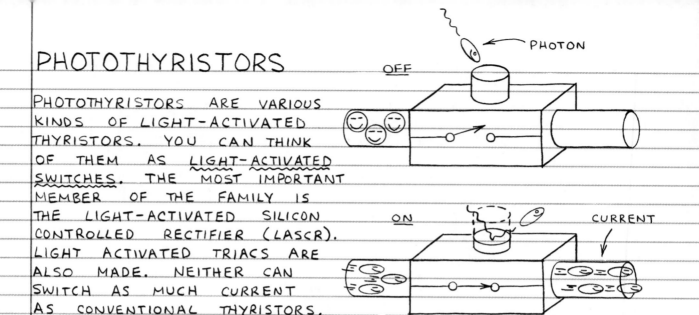

LIGHT ACTIVATED SCRs (LASCRs)

TO IMPROVE THEIR SENSITIVITY TO LIGHT, THE LASCR IS MADE THINNER THAN STANDARD SCRs. THIS LIMITS THE AMOUNT OF CURRENT THEY CAN SWITCH. FOR HIGH CURRENT APPLICATIONS A LASCR CAN BE USED TO TRIGGER A CONVENTIONAL SCR.

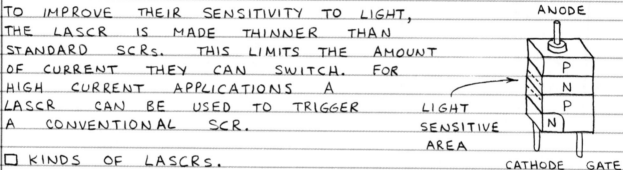

☐ KINDS OF LASCRs.

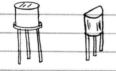

MOST LASCRs CAN SWITCH UP TO A FEW HUNDRED VOLTS. MAXIMUM CURRENT IS ONLY A FEW TENTHS OF AN AMPERE.

HOW LASCRs ARE USED

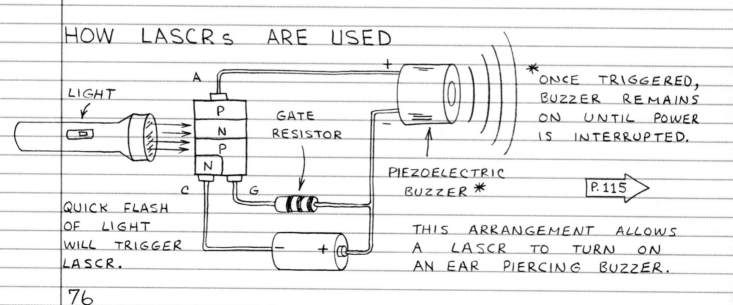

* ONCE TRIGGERED, BUZZER REMAINS ON UNTIL POWER IS INTERRUPTED.

QUICK FLASH OF LIGHT WILL TRIGGER LASCR.

THIS ARRANGEMENT ALLOWS A LASCR TO TURN ON AN EAR PIERCING BUZZER.

P.115 →

SOLAR CELLS

SOLAR CELLS ARE PN JUNCTION PHOTODIODES WITH AN EXCEPTIONALLY LARGE LIGHT SENSITIVE AREA. A SINGLE SILICON SOLAR CELL GENERATES 0.5 VOLT IN BRIGHT SUNLIGHT.

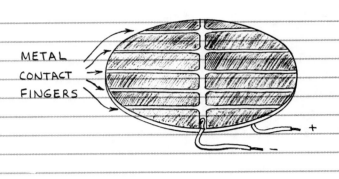
METAL CONTACT FINGERS

☐ SOLAR CELL OPERATION

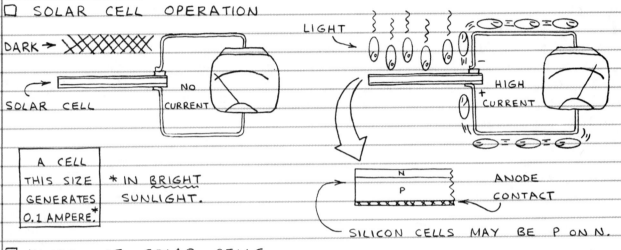

A CELL THIS SIZE GENERATES 0.1 AMPERE.*

*IN BRIGHT SUNLIGHT.

SILICON CELLS MAY BE P ON N.

☐ KINDS OF SOLAR CELLS.

MANY DIFFERENT KINDS OF SILICON SOLAR CELLS ARE MADE. OFTEN INDIVIDUAL CELLS ARE CONNECTED IN SERIES OR PARALLEL.

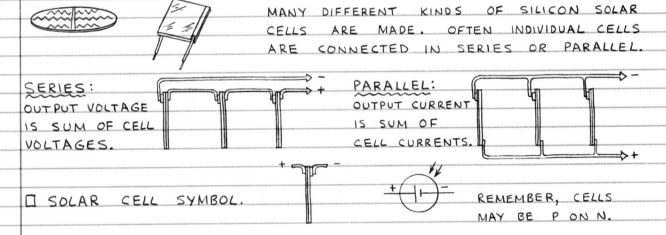

SERIES: OUTPUT VOLTAGE IS SUM OF CELL VOLTAGES.

PARALLEL: OUTPUT CURRENT IS SUM OF CELL CURRENTS.

☐ SOLAR CELL SYMBOL.

REMEMBER, CELLS MAY BE P ON N.

HOW SOLAR CELLS ARE USED

P. 115 →

ARRAYS OF SOLAR CELLS CAN CHARGE RECHARGEABLE CELLS AND BATTERIES.

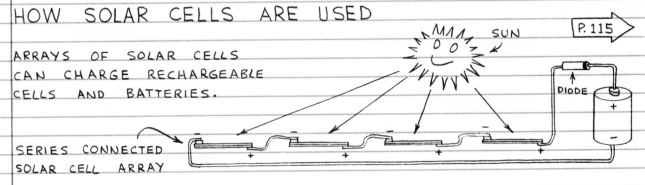

SERIES CONNECTED SOLAR CELL ARRAY

77

5. INTEGRATED CIRCUITS

Electronic circuits can be made by simultaneously forming individual transistors, diodes and resistors on a small chip of silicon. The components are connected to one another with aluminum "wires" deposited on the surface of the chip. The result is an <u>integrated circuit</u>.

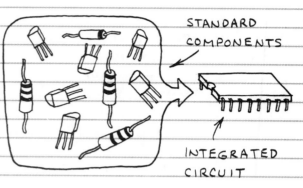

Integrated circuits (or IC's) can contain as few as several to as many as hundreds of thousands of transistors. They have made possible video games, digital watches, affordable computers and many other very sophisticated products. Here's a simplified and highly magnified view of a section of a bipolar integrated circuit:

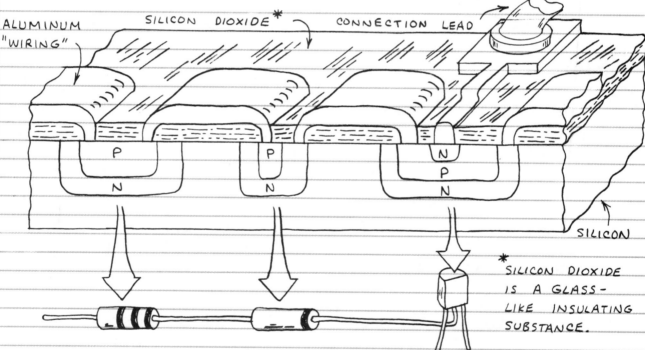

* Silicon dioxide is a glass-like insulating substance.

RESISTOR — A small section of P-type silicon forms a resistor.

DIODE — A PN junction forms a diode.

TRANSISTOR — A pair of PN junctions forms an NPN transistor.

Of course the conventional components shown below the highly magnified section of the IC are <u>not</u> drawn to the same scale. For example, one kind of IC includes 262,144 transistors on a silicon chip only about 1/4 inch square!

☐ KINDS OF INTEGRATED CIRCUITS — INTEGRATED CIRCUITS ARE GROUPED INTO TWO MAJOR CATEGORIES:

1. ANALOG (OR LINEAR) IC'S PRODUCE, AMPLIFY OR RESPOND TO VARIABLE VOLTAGES. ANALOG IC'S INCLUDE MANY KINDS OF AMPLIFIERS, TIMERS, OSCILLATORS AND VOLTAGE REGULATORS.

2. DIGITAL (OR LOGIC) IC'S RESPOND TO OR PRODUCE SIGNALS HAVING ONLY TWO VOLTAGE LEVELS. DIGITAL IC'S INCLUDE MICROPROCESSORS, MEMORIES, MICROCOMPUTERS AND MANY KINDS OF SIMPLER CHIPS.

SOME IC'S COMBINE ANALOG AND DIGITAL FUNCTIONS ON A SINGLE CHIP. FOR EXAMPLE, A DIGITAL CHIP MAY INCLUDE A BUILT-IN ANALOG VOLTAGE REGULATOR SECTION. AND AN ANALOG TIMER CHIP MAY INCLUDE AN ON-CHIP DIGITAL COUNTER TO GIVE MUCH LONGER TIME DELAYS THAN POSSIBLE WITH THE TIMER ALONE.

VOLTAGE INTO OR OUT OF CHIP. (NOT POWER SUPPLY VOLTAGE.)

ANALOG IC'S — TIME

DIGITAL IC'S — TIME

☐ KINDS OF INTEGRATED CIRCUIT PACKAGES — IC CHIPS ARE SUPPLIED IN MANY DIFFERENT PACKAGES. BY FAR THE MOST COMMON ARE VARIATIONS OF THE DUAL IN-LINE PACKAGE (OR DIP). THE DIP IS MADE FROM PLASTIC (CHEAP) OR CERAMIC (MORE ROBUST). MOST DIPS HAVE 14 OR 16 PINS, BUT THE PIN COUNT CAN RANGE FROM 4 TO 64. HERE'S A TYPICAL DIP:

INDEX MARKER (INDICATES PIN 1)

MANUFACTURER'S LOGO (MOTOROLA)

PART NUMBER — MC14021B

DATE CODE — CP8314
83 = 1983
14 = 14th WEEK

PIN NUMBERS

ANOTHER IC PACKAGE IS THE TO-5 METAL CAN. THOUGH VERY STURDY, IT'S BEING REPLACED IN MANY CASES BY CHEAPER PLASTIC DIPS.

METAL CAN

LEADS

79

6. DIGITAL INTEGRATED CIRCUITS

NO MATTER HOW COMPLICATED, ALL DIGITAL INTEGRATED CIRCUITS ARE MADE FROM SIMPLE BUILDING BLOCKS CALLED GATES. GATES ARE LIKE ELECTRONICALLY CONTROLLED SWITCHES. THEY ARE EITHER ON OR OFF. HOW DO GATES WORK? LET'S START WITH THE BASICS...

MECHANICAL SWITCH GATES

THE THREE SIMPLEST GATES CAN BE DEMONSTRATED WITH SOME PUSHBUTTON SWITCHES, A BATTERY AND A LAMP.

☐ SWITCH "AND" GATE. THE LAMP GLOWS ONLY WHEN SWITCHES A AND B ARE CLOSED. THE TABLE SUMMARIZES THE GATE'S OPERATION. IT'S CALLED A TRUTH TABLE.

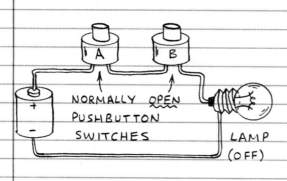

	A	B	OUT
OPEN SWITCH = OFF	OFF	OFF	OFF
CLOSED SWITCH = ON	OFF	ON	OFF
	ON	OFF	OFF
ALL POSSIBLE ON-OFF COMBINATIONS	ON	ON	ON

☐ SWITCH "OR" GATE. THE LAMP GLOWS ONLY WHEN SWITCH A OR SWITCH B OR BOTH SWITCHES A AND B ARE CLOSED. HERE'S THE TRUTH TABLE:

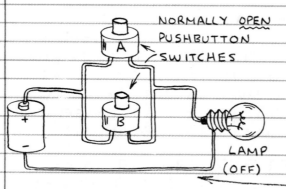

	A	B	OUT
THE SWITCHES ARE THE GATE'S	OFF	OFF	OFF
INPUTS. THE LEAD	OFF	ON	ON
WITHOUT SWITCHES	ON	OFF	ON
IS THE COMMON OR GROUND LEAD.	ON	ON	ON

☐ SWITCH "NOT" GATE. THE LAMP NORMALLY GLOWS. ONLY WHEN THE SWITCH IS OPENED IS THE LAMP OFF. IN OTHER WORDS, THE "NOT" GATE REVERSES (INVERTS) THE USUAL ACTION OF A SWITCH. HERE'S THE TRUTH TABLE:

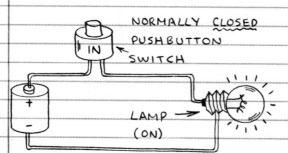

	IN	OUT
THE "NOT" GATE IS USUALLY CALLED THE INVERTER.	OFF	ON
	ON	OFF

THE BINARY CONNECTION

IT'S POSSIBLE TO SUBSTITUTE THE DIGITS 0 AND 1 FOR THE OFF AND ON STATES OF A SWITCH. THE TRUTH TABLES FOR THE GATES ON THE PREVIOUS PAGE THEN BECOME:

"AND" GATE

A	B	OUT
0	0	0
0	1	0
1	0	0
1	1	1

"OR" GATE

A	B	OUT
0	0	0
0	1	1
1	0	1
1	1	1

"NOT" GATE

IN	OUT
0	1
1	0

THE 0 AND 1 INPUT (A & B) COMBINATIONS FORM NUMBERS IN THE TWO DIGIT (OR BIT) BINARY NUMBER SYSTEM. IN DIGITAL ELECTRONICS, BINARY NUMBERS SERVE AS CODES THAT REPRESENT DECIMAL NUMBERS, LETTERS OF THE ALPHABET, VOLTAGES AND MANY OTHER KINDS OF INFORMATION.

DECIMAL	BINARY	BINARY-CODED DECIMAL (BCD)
0	0	0000 0000
1	1	0000 0001
2	10	0000 0010
3	11	0000 0011
4	100	0000 0100
5	101	0000 0101
6	110	0000 0110
7	111	0000 0111
8	1000	0000 1000
9	1001	0000 1001
10	1010	0001 0000
11	1011	0001 0001
12	1100	0001 0010
13	1101	0001 0011
14	1110	0001 0100
15	1111	0001 0101

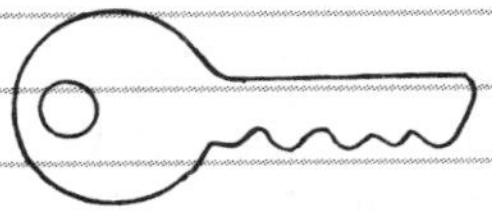

BINARY FACTS

A BINARY 0 OR 1 IS A BIT.
A PATTERN OF 4 BITS IS A NIBBLE.
A PATTERN OF 8 BITS IS A BYTE.

BCD— EACH DECIMAL DIGIT IS ASSIGNED ITS BINARY EQUIVALENT. NOTE THAT LEADING ZEROS ARE SHOWN. IN DIGITAL ELECTRONICS ALL BIT LOCATIONS ARE OCCUPIED.

BINARY NUMBERS CAN BE SENT THROUGH WIRES (BUSES) ALL AT ONCE (PARALLEL) OR A BIT AT A TIME (SERIAL). SHOWN HERE ARE SERIAL AND PARALLEL TRANSMISSION OF 15...14...13...12.

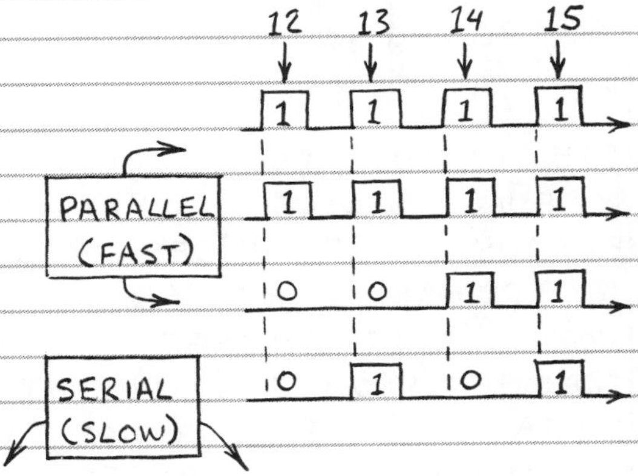

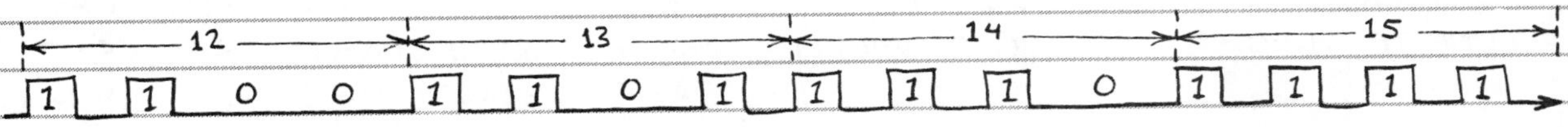

DIODE GATES

Often it's desirable to control a gate electrically rather than mechanically. The simplest electrically controlled gate uses PN junction diodes that are switched ON (forward bias) or OFF (reverse bias) by an input signal of several volts (binary 1 or HIGH) or an input near or at ground (binary 0 or LOW).

☐ DIODE "OR" GATE   ☐ DIODE "AND" GATE

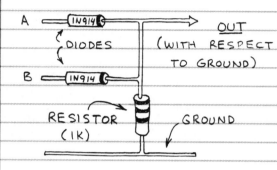
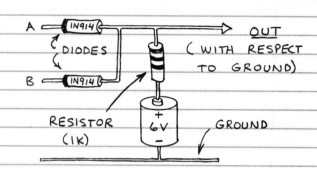

When the input voltage at A or B is more positive than ground, it passes through the forward biased diode(s) and appears at the output. Otherwise the output is at or near ground. The truth table is valid for inputs of 0 volt (0 or LOW) and +6 volts (1 or HIGH).

When the input voltage at A and B is more positive than ground, current flows from the battery through the resistor to the output. If either A or B is at or near ground, one or both diodes become forward biased and current flows away from the output.

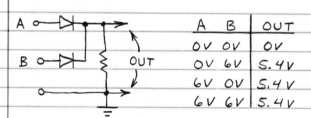

A	B	OUT
0V	0V	0V
0V	6V	5.4V
6V	0V	5.4V
6V	6V	5.4V

A	B	OUT
0V	0V	0V
0V	6V	.5V
6V	0V	.5V
6V	6V	5.4V

The output does not reach a full 6 volts when high because the diodes require a forward voltage of 0.6 volt. This voltage is subtracted from the output voltage. (In electronics jargon a silicon diode causes a "voltage drop" of 0.6 volt.)

As circuits become more complicated, pictorial views are not practical. That's why this page introduces circuit diagrams for each of the two pictorials shown above. We'll find out more about circuit diagrams later. In the meantime, the next page shows more of them...

TRANSISTOR GATES

THE VOLTAGE DROP OF DIODE GATES MEANS AMPLIFICATION IS REQUIRED IN ORDER TO CONNECT TOGETHER A SERIES OF GATES. WHILE TRANSISTORS CAN PROVIDE THE NECESSARY AMPLIFICATION, TRANSISTORS CAN FUNCTION AS GATES! BOTH BIPOLAR AND FIELD-EFFECT TRANSISTORS CAN BE USED. ON THIS PAGE ARE SHOWN CIRCUIT DIAGRAMS FOR SOME OF THE SIMPLEST BIPOLAR TRANSISTOR GATES. TOGETHER THEY FORM THE RESISTOR-TRANSISTOR DIGITAL LOGIC FAMILY. YOU CAN ACTUALLY MAKE THESE GATES. BUT THE MAIN REASON THEY'RE HERE IS TO GIVE YOU AN APPRECIATION FOR THE INTEGRATED CIRCUIT GATES WE'LL BE LOOKING AT SHORTLY....

☐ "NOT" GATE (INVERTER)

WHEN IN IS AT +V (BINARY 1 OR HIGH), TRANSISTOR Q1 SWITCHES ON AND CONNECTS OUT DIRECTLY TO GROUND (BINARY 0 OR LOW). WHEN IN IS LOW, Q1 SWITCHES OFF AND OUT BECOMES (THROUGH R1) +V. "NOT" GATES LIKE THIS MAKE POSSIBLE IMPORTANT NEW LOGIC GATES.

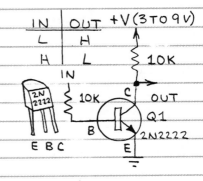

IN	OUT
L	H
H	L

☐ "AND" GATE   ☐ "NAND" (NOT-AND) GATE

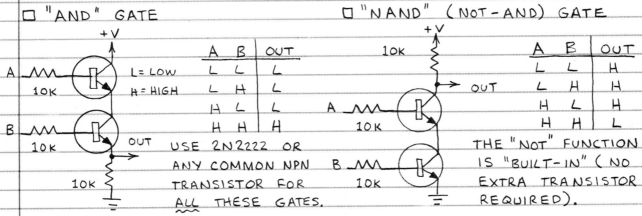

A	B	OUT
L	L	L
L	H	L
H	L	L
H	H	H

L = LOW
H = HIGH

USE 2N2222 OR ANY COMMON NPN TRANSISTOR FOR ALL THESE GATES.

A	B	OUT
L	L	H
L	H	H
H	L	H
H	H	L

THE "NOT" FUNCTION IS "BUILT-IN" (NO EXTRA TRANSISTOR REQUIRED).

☐ "OR" GATE   ☐ "NOR" (NOT-OR) GATE

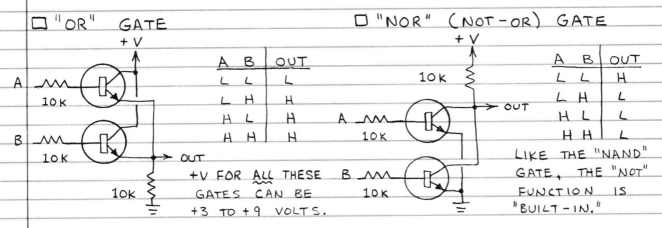

A	B	OUT
L	L	L
L	H	H
H	L	H
H	H	H

+V FOR ALL THESE GATES CAN BE +3 TO +9 VOLTS.

A	B	OUT
L	L	H
L	H	L
H	L	L
H	H	L

LIKE THE "NAND" GATE, THE "NOT" FUNCTION IS "BUILT-IN."

GATE SYMBOLS

BEFORE MOVING ON TO DIGITAL INTEGRATED CIRCUITS, LET'S LOOK AT THE SYMBOLS FOR THE VARIOUS KINDS OF GATES. THIS IS ALSO A GOOD PLACE TO INTRODUCE SEVERAL GATES WE'VE NOT YET ENCOUNTERED.

☐ "AND" GATE

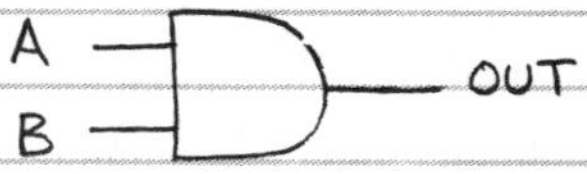

A	B	OUT
L	L	L
L	H	L
H	L	L
H	H	H

☐ "NAND" GATE

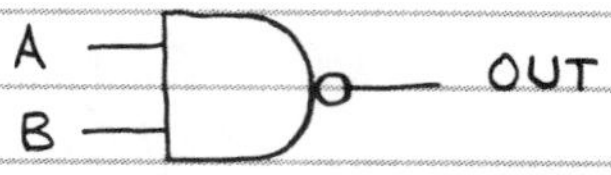

A	B	OUT
L	L	H
L	H	H
H	L	H
H	H	L

☐ "OR" GATE

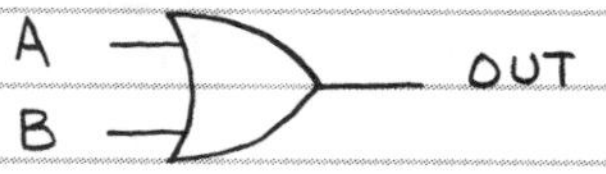

A	B	OUT
L	L	L
L	H	H
H	L	H
H	H	H

☐ "NOR" GATE

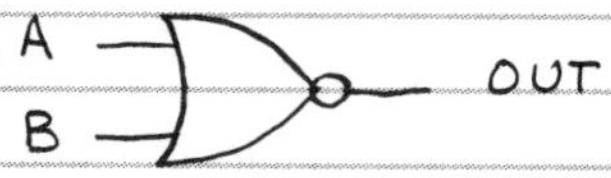

A	B	OUT
L	L	H
L	H	L
H	L	L
H	H	L

☐ "EXCLUSIVE-OR" GATE

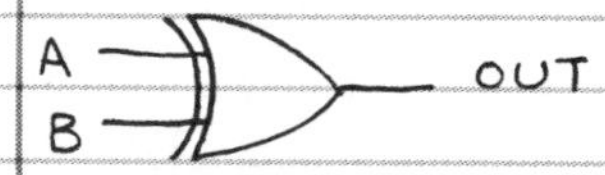

A	B	OUT
L	L	L
L	H	H
H	L	H
H	H	L

☐ "EXCLUSIVE-NOR" GATE

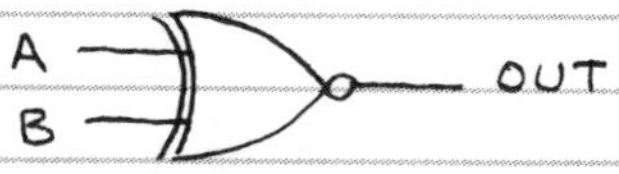

A	B	OUT
L	L	H
L	H	L
H	L	L
H	H	H

☐ LOGIC GATES WITH MORE THAN TWO INPUTS — THE GATES SHOWN ABOVE ARE CALLED LOGIC CIRCUITS BECAUSE THEY MAKE LOGICAL DECISIONS. LOGIC GATES OFTEN HAVE MORE THAN TWO INPUTS. ADDITIONAL INPUTS INCREASE THE DECISION MAKING POWER OF A GATE. THEY ALSO INCREASE THE NUMBER OF WAYS GATES CAN BE CONNECTED TO ONE ANOTHER TO FORM ADVANCED DIGITAL LOGIC CIRCUITS. HERE ARE TWO EXAMPLES:

3-INPUT "AND" GATE

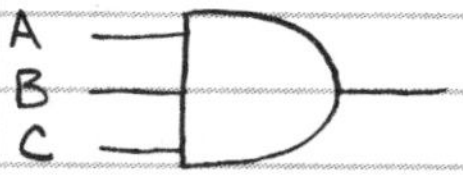

A	B	C	OUT
L	L	L	L
L	L	H	L
L	H	L	L
L	H	H	L
H	L	L	L
H	L	H	L
H	H	L	L
H	H	H	H

3-INPUT "NAND" GATE

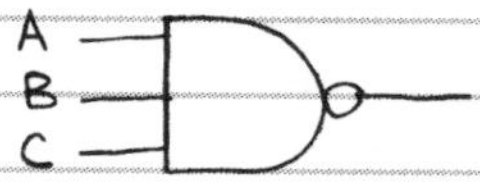

A	B	C	OUT
L	L	L	H
L	L	H	H
L	H	L	H
L	H	H	H
H	L	L	H
H	L	H	H
H	H	L	H
H	H	H	L

☐ **SINGLE-INPUT GATES —** THE "NOT" GATE OR INVERTER IS VERY IMPORTANT SINCE IT CAN INVERT (REVERSE) THE OUTPUT FROM ANOTHER GATE. STRICTLY SPEAKING, HOWEVER, THE INVERTER IS NOT A DECISION MAKING CIRCUIT (LIKE GATES WITH TWO OR MORE INPUTS). A CLOSE RELATIVE OF THE INVERTER IS THE BUFFER, A NON-INVERTING CIRCUIT THAT ISOLATES GATES FROM OTHER CIRCUITS OR ALLOWS THEM TO DRIVE HIGHER THAN NORMAL LOADS. THREE-STATE INVERTERS AND BUFFERS HAVE AN OUTPUT THAT CAN BE ELECTRONICALLY DISCONNECTED FROM THE REMAINDER OF THE CIRCUIT. THE OUTPUT IS THEN NEITHER HIGH NOR LOW. INSTEAD IT "FLOATS" AND APPEARS AS A VERY HIGH RESISTANCE.

☐ **BUFFER**

IN	OUT
L	L
H	H

☐ **INVERTER ("NOT" GATE)**

IN	OUT
L	H
H	L

☐ **3-STATE BUFFER**

CONTROL	IN	OUT
L	L	L
L	H	H
H	X	HI-Z

☐ **3-STATE INVERTER**

CONTROL	IN	OUT
L	L	H
L	H	L
H	X	HI-Z

"X" MEANS "DOESN'T MATTER." HI-Z MEANS HIGH OUTPUT RESISTANCE.

DATA "HIGHWAYS"

OFTEN CIRCUITS MADE FROM GATES EXCHANGE INFORMATION (BINARY 0'S AND 1'S ENCODED AS LOW AND HIGH VOLTAGE LEVELS). THE INFORMATION IS USUALLY SENT OVER WIRES CALLED BUSES. A BUS IS LIKE A DATA HIGHWAY. IT MAY BE ONE WIRE THROUGH WHICH INFORMATION IS SENT SERIALLY (BIT BY BIT). OR IT MAY BE UP TO EIGHT (OR MORE) WIRES THROUGH WHICH INFORMATION IS SENT IN PARALLEL (A BYTE OR MORE AT A TIME). IN BOTH CASES, OF COURSE, A GROUND IS REQUIRED TO COMPLETE THE CIRCUIT.

☐ **3-STATE TRAFFIC COPS —** 3-STATE GATES CAN STOP "TRAFFIC JAMS" ON BUSES. FOR INSTANCE:

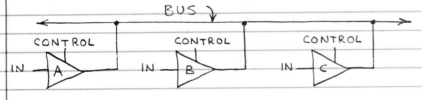

ONLY DATA ENTERING THE SELECTED BUFFER (CONTROL = L) GETS ON THE BUS.

HOW GATES ARE USED

Gates can be used individually or connected together to form a "network" of gates called a LOGIC CIRCUIT. Almost all logic circuits can be placed in one of two categories: COMBINATIONAL or SEQUENTIAL.

COMBINATIONAL LOGIC CIRCUITS

Combinational logic circuits respond to incoming DATA (0's and 1's) almost immediately and without regard to earlier events. (This will make more sense when you read about sequential circuits...) Combinational logic circuits can be very simple or immensely complicated. Virtually ANY combinational circuit can be implemented with only "NAND" or "NOR" gates. Like these "NAND" gate circuits...

NOTE: These circuits do not show the ground connection that must be present. Usually the ground is common to the input and output.

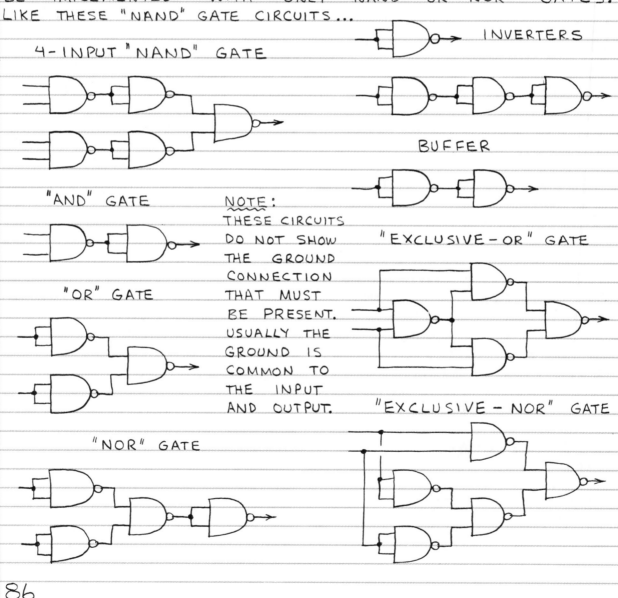

☐ **COMBINING DIFFERENT GATES** — HERE ARE TWO EXAMPLES OF <u>COMBINATIONAL</u> NETWORKS THAT USE MORE THAN ONE KIND OF GATE. (REMEMBER, BOTH THESE CIRCUITS CAN ALSO BE MADE ENTIRELY FROM "NAND" GATES!)

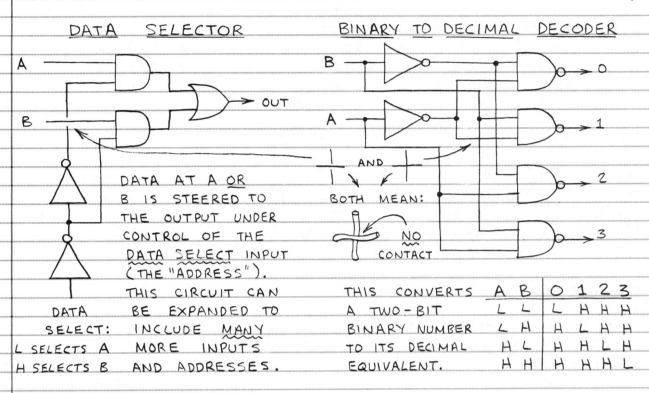

DATA SELECTOR

DATA AT A <u>OR</u> B IS STEERED TO THE OUTPUT UNDER CONTROL OF THE <u>DATA SELECT</u> INPUT (THE "ADDRESS").

DATA SELECT:
L SELECTS A
H SELECTS B

THIS CIRCUIT CAN BE EXPANDED TO INCLUDE <u>MANY</u> MORE INPUTS AND ADDRESSES.

BINARY TO DECIMAL DECODER

BOTH MEAN: NO CONTACT

THIS CONVERTS A TWO-BIT BINARY NUMBER TO ITS DECIMAL EQUIVALENT.

A	B	0	1	2	3
L	L	L	H	H	H
L	H	H	L	H	H
H	L	H	H	L	H
H	H	H	H	H	L

☐ **ADVANCED COMBINATIONAL NETWORKS** — HERE ARE SOME SIMPLE EXAMPLES FROM FOUR MAJOR FAMILIES OF COMBINATIONAL NETWORKS. THESE AND MANY OTHER NETWORK FAMILIES ARE AVAILABLE AS <u>INTEGRATED CIRCUITS</u>. BOXES LIKE THOSE SHOWN ARE LOGIC CIRCUIT SYMBOLS THAT REPRESENT COMPLICATED NETWORKS OF GATES.

MULTIPLEXER (DATA SELECTOR)

X	Y	OUT
L	L	A
L	H	B
H	L	C
H	H	D

DEMULTIPLEXER

X	Y	IN TO...
L	L	A
L	H	B
H	L	C
H	H	D

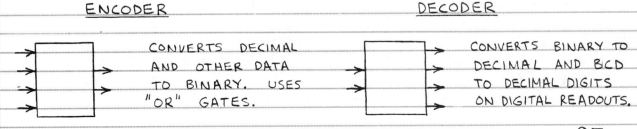

ENCODER

CONVERTS DECIMAL AND OTHER DATA TO BINARY. USES "OR" GATES.

DECODER

CONVERTS BINARY TO DECIMAL AND BCD TO DECIMAL DIGITS ON DIGITAL READOUTS.

SEQUENTIAL LOGIC CIRCUITS

THE OUTPUT STATE OF A SEQUENTIAL LOGIC CIRCUIT IS DETERMINED BY THE PREVIOUS STATE OF THE INPUT. IN OTHER WORDS, BITS OF DATA MOVE THROUGH SEQUENTIAL CIRCUITS STEP-BY-STEP. OFTEN THE DATA ADVANCES ONE STEP WHEN A PULSE IS RECEIVED FROM A "CLOCK" (A CIRCUIT THAT EMITS A STEADY STREAM OF PULSES). THE SEQUENTIAL LOGIC BUILDING BLOCK IS THE <u>FLIP-FLOP</u>. HERE'S A QUICK FLIP-FLOP REVIEW:

☐ THE BASIC "RS" (<u>R</u>ESET-<u>S</u>ET) FLIP-FLOP

ALSO CALLED A <u>LATCH</u>. THE OUTPUTS (Q AND $\bar{Q}$) ARE ALWAYS IN OPPOSITE STATES. ($\bar{Q}$ MEANS "NOT Q".)

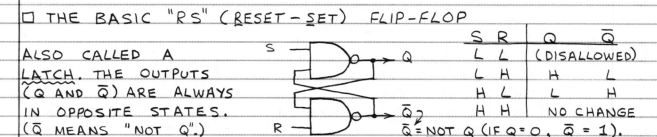

S	R	Q	$\bar{Q}$
L	L	(DISALLOWED)	
L	H	H	L
H	L	L	H
H	H	NO CHANGE	

$\bar{Q}$ = <u>NOT</u> Q (IF Q=0, $\bar{Q}$=1).

☐ CLOCKED "RS" FLIP-FLOP

THIS LATCH IGNORES DATA AT S AND R UNTIL A "CLOCK" (OR <u>ENABLE</u>) PULSE ARRIVES. THEN IT CHANGES STATES.

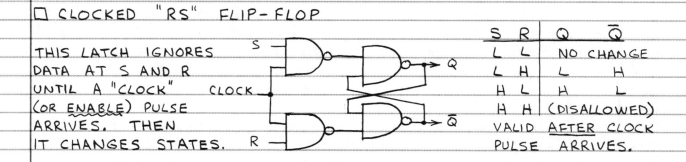

S	R	Q	$\bar{Q}$
L	L	NO CHANGE	
L	H	L	H
H	L	H	L
H	H	(DISALLOWED)	

VALID <u>AFTER</u> CLOCK PULSE ARRIVES.

☐ "D" (<u>D</u>ATA OR <u>D</u>ELAY) FLIP-FLOP

D FLIP-FLOPS <u>STORE</u> THE CURRENT OUTPUTS BETWEEN CLOCK PULSES.

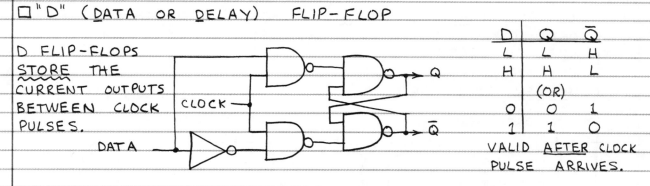

D	Q	$\bar{Q}$
L	L	H
H	H	L
(OR)		
0	0	1
1	1	0

VALID <u>AFTER</u> CLOCK PULSE ARRIVES.

☐ "JK" FLIP-FLOP

THE "JK" FLIP-FLOP ALLOWS BOTH INPUTS TO BE H. (IN WHICH CASE ITS OUTPUTS "TOGGLE" OR SWITCH STATES AT EACH CLOCK PULSE.)

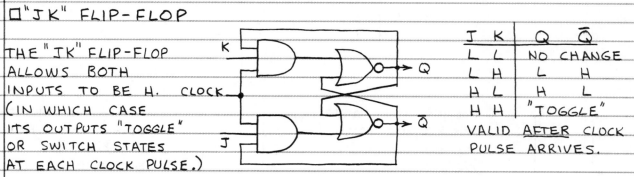

J	K	Q	$\bar{Q}$
L	L	NO CHANGE	
L	H	L	H
H	L	H	L
H	H	"TOGGLE"	

VALID <u>AFTER</u> CLOCK PULSE ARRIVES.

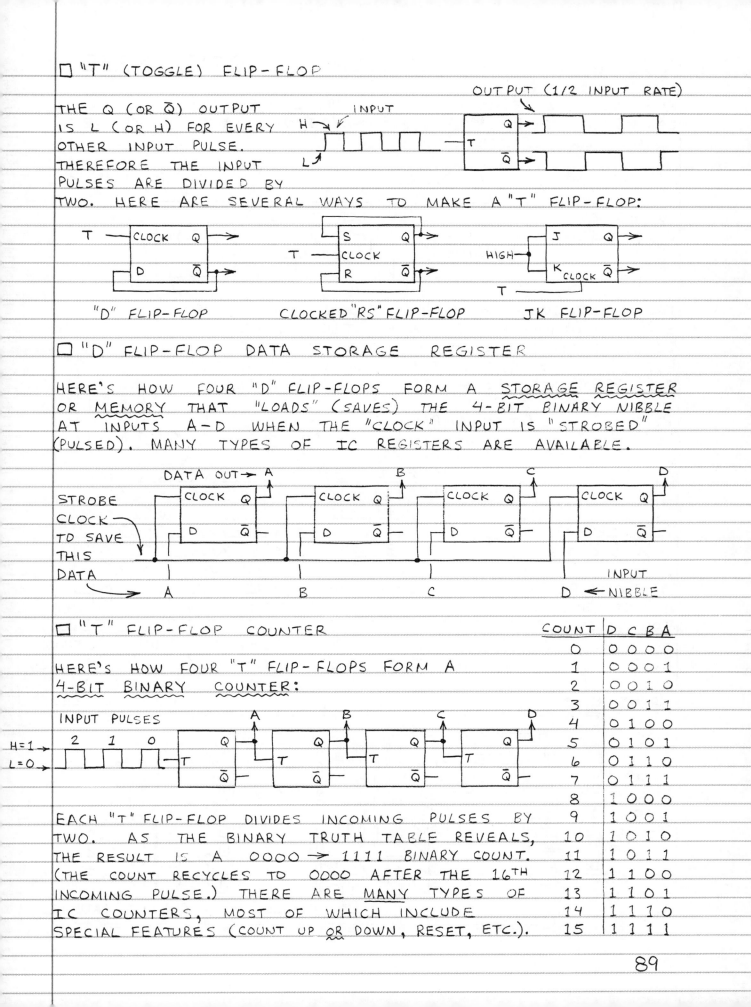

A COMBINATIONAL–SEQUENTIAL LOGIC SYSTEM

HERE'S HOW COMBINATIONAL AND SEQUENTIAL LOGIC IC'S CAN FORM A <u>DECIMAL COUNTING CIRCUIT</u>, A VERY SIMPLE DIGITAL LOGIC SYSTEM.

1. THE BLOCK DIAGRAM → P. 117

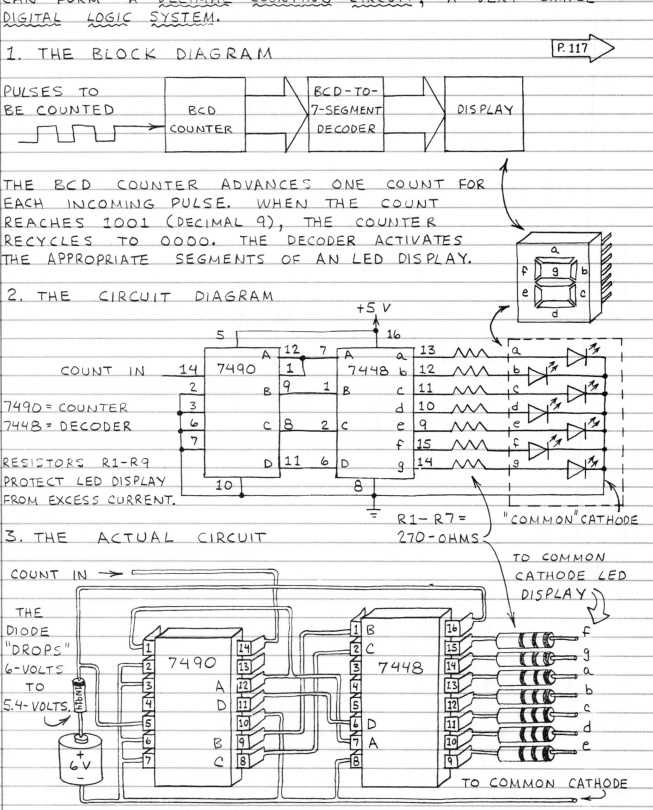

THE BCD COUNTER ADVANCES ONE COUNT FOR EACH INCOMING PULSE. WHEN THE COUNT REACHES 1001 (DECIMAL 9), THE COUNTER RECYCLES TO 0000. THE DECODER ACTIVATES THE APPROPRIATE SEGMENTS OF AN LED DISPLAY.

2. THE CIRCUIT DIAGRAM

7490 = COUNTER
7448 = DECODER

RESISTORS R1–R9 PROTECT LED DISPLAY FROM EXCESS CURRENT.

R1–R7 = 270-OHMS

"COMMON" CATHODE

3. THE ACTUAL CIRCUIT

THE DIODE "DROPS" 6-VOLTS TO 5.4-VOLTS.

TO COMMON CATHODE LED DISPLAY

TO COMMON CATHODE

DIGITAL IC FAMILIES

There are more than a dozen major families of bipolar and MOS integrated circuits. Each IC (or "chip") contains a specific logic network or assortment of various logic functions. Here are some of the major digital IC families:

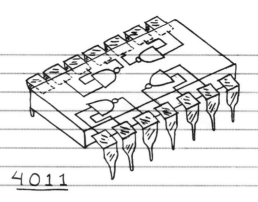

4011

A "QUAD" (four) of 2-input CMOS "NAND" gates

☐ BIPOLAR DIGITAL IC's

1. TRANSISTOR-TRANSISTOR LOGIC (TTL OR T^2L). The largest and formerly most popular digital IC family. Can change states more than 20,000,000 times per second. Very inexpensive. Drawbacks: must be powered by 5-volt supply. Uses LOTS OF POWER. (Individual gates require 3 or 4 milliamperes.) Most widely used is the 7400 series. The 7404, for example, contains four inverters.

2. LOW-POWER SCHOTTKY TTL (LS). A newer kind of TTL that consumes only 20% as much power. Drawback: more expensive than standard TTL. Most widely used is the 74LS00 series.

☐ MOSFET DIGITAL IC's

1. P- AND N-CHANNEL MOS (PMOS AND NMOS). Contain more gates per chip than TTL. Many special purpose chips (microprocessors, memories, etc.). Drawbacks: few counterparts to popular TTL chips. Slower than TTL. May require two or more supply voltages. May be damaged by static electrical discharge.

2. COMPLEMENTARY MOS (CMOS). Fastest growing and most versatile digital IC family. There are CMOS versions of most popular TTL chips. One series uses the same designation numbers. The 74C04, for example, is the CMOS equivalent of the TTL 7404. New high-speed CMOS just as fast as TTL. Most CMOS has a wide supply voltage range (typically +3 to +18 volts). Uses less power than any other digital IC family. (Individual gates require 0.1 milliampere.) Drawback: may be damaged by static electrical discharge. Most widely used are 74C00 and 4000 series.

7. LINEAR INTEGRATED CIRCUITS

THE INPUT AND OUTPUT VOLTAGE LEVELS OF LINEAR INTEGRATED CIRCUITS CAN VARY OVER A <u>WIDE</u> RANGE. OFTEN THE OUTPUT VOLTAGE IS PROPORTIONAL TO THE INPUT VOLTAGE. THEREFORE, A GRAPH OF INPUT VERSUS OUTPUT IS A STRAIGHT (<u>LINEAR</u>) LINE. THERE ARE MANY TYPES OF LINEAR IC'S. ONLY THE MAJOR TYPES ARE COVERED HERE. FIRST LET'S COMPARE THE BASIC DIGITAL AND LINEAR CIRCUITS:

THE BASIC LINEAR CIRCUIT

A SINGLE BIPOLAR OR FIELD-EFFECT TRANSISTOR CAN FUNCTION AS A DIGITAL OR LINEAR CIRCUIT. IN BOTH CASES, THE TRANSISTOR CAN <u>INVERT</u> THE SIGNAL AT ITS INPUT. HERE'S HOW AN NPN BIPOLAR TRANSISTOR CAN PERFORM ALL FOUR FUNCTIONS:

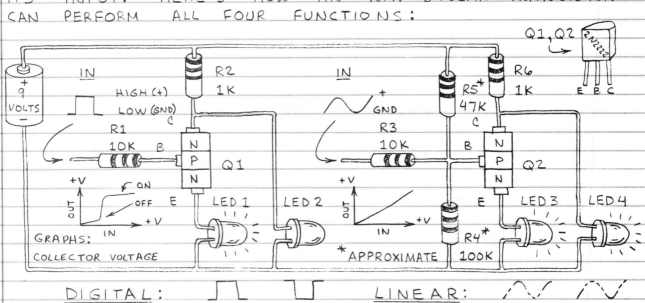

HERE TRANSISTOR Q1 IS USED AS A <u>SWITCH</u>. WHEN THE INPUT IS NEAR +V (OR HIGH), Q1 TURNS <u>ON</u> AND LED 1 IS ILLUMINATED. WHEN THE INPUT IS NEAR GROUND (OR LOW), Q1 TURNS <u>OFF</u>. THIS TURNS LED 1 <u>OFF</u> AND ALLOWS LED 2 TO GLOW. (R2 CONTROLS THE CURRENT THROUGH BOTH LEDs.) THIS CIRCUIT IS THEN A COMBINED DIGITAL <u>BUFFER</u> AND <u>INVERTER</u>.

HERE Q2 IS AN AMPLIFIER THAT OPERATES OVER THE ENTIRE RANGE FROM FULL <u>OFF</u> TO FULL <u>ON</u>. R4 AND R5 FORM A VOLTAGE DIVIDER THAT APPLIES A SMALL VOLTAGE TO Q2'S BASE TO KEEP Q2 SLIGHTLY ON EVEN WHEN NO INPUT IS PRESENT. THIS ALLOWS Q2 TO OPERATE IN A <u>LINEAR</u> MODE. AS THE INPUT VOLTAGE RISES, LED 3 BRIGHTENS AND LED 4 DIMS.

OPERATIONAL AMPLIFIERS

OPERATIONAL AMPLIFIERS (OR "OP-AMPS") ARE BY FAR THE MOST VERSATILE OF LINEAR IC'S. THEY'RE CALLED "OPERATIONAL" AMPLIFIERS SINCE THEY WERE ORIGINALLY DESIGNED TO DO MATHEMATICAL OPERATIONS. OP-AMPS AMPLIFY THE DIFFERENCE BETWEEN VOLTAGES OR SIGNALS (AC OR DC) APPLIED TO THEIR TWO INPUTS. THE VOLTAGE APPLIED TO ONLY ONE INPUT WILL BE AMPLIFIED IF THE SECOND INPUT IS GROUNDED OR MAINTAINED AT SOME VOLTAGE LEVEL.

☐ OP-AMP OPERATION — THE OP-AMP HAS AN INVERTING AND NON-INVERTING INPUT. THE POLARITY OF A VOLTAGE APPLIED TO THE INVERTING INPUT IS REVERSED AT THE OUTPUT. (INVERTING INPUT IS − ; NON-INVERTING INPUT IS +.)

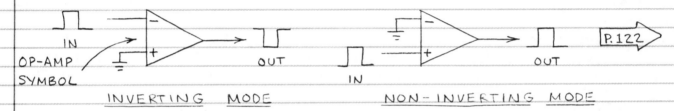

OP-AMP SYMBOL   INVERTING MODE   NON-INVERTING MODE   P.122 →

☐ OP-AMP "FEEDBACK" — THE CIRCUITS SHOWN ABOVE ALLOW THE OP-AMP TO OPERATE AT ITS MAXIMUM AMPLIFICATION LEVEL (OR GAIN). USUALLY THE GAIN IS REDUCED TO A MORE PRACTICAL LEVEL BY FEEDING SOME OF THE OUTPUT BACK TO THE INVERTING (−) INPUT. FOR EXAMPLE:

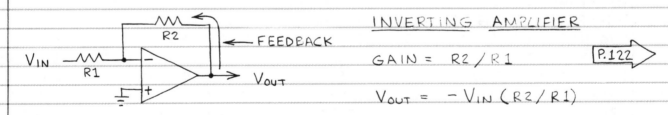

INVERTING AMPLIFIER

GAIN = R2/R1

$V_{OUT} = -V_{IN}(R2/R1)$

P.122 →

☐ OP-AMP COMPARATOR — WHEN OPERATED WITHOUT A FEEDBACK RESISTOR (R2 ABOVE), THE OUTPUT VOLTAGE WILL SWING FROM FULL ON TO FULL OFF (OR VICE VERSA) WHEN THE VOLTAGES APPLIED TO THE INPUTS DIFFER BY ONLY ABOUT 0.001 VOLT! THIS DIGITAL-LIKE MODE MAKES POSSIBLE MANY USEFUL APPLICATIONS. P.124 →

☐ TYPES OF OP-AMPS — BOTH BIPOLAR AND MOSFET IC OP-AMPS ARE AVAILABLE. SOME BIPOLAR OP-AMPS HAVE FET OR MOSFET INPUTS TO PROVIDE VERY HIGH INPUT RESISTANCE. MANY DIFFERENT OP-AMPS ARE MADE. A SINGLE IC MAY INCLUDE UP TO FOUR INDIVIDUAL OP-AMPS.

TIMERS

WHEN OPERATED AS A COMPARATOR, THE OP-AMP CAN BE USED AS A TIMER. ALL THAT'S REQUIRED IS AN RC (RESISTOR-CAPACITOR) CIRCUIT LIKE THIS:

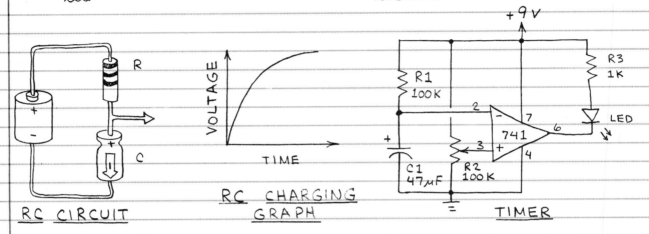

RC CIRCUIT — **RC CHARGING GRAPH** — **TIMER**

IN THE CIRCUIT DIAGRAM (ABOVE RIGHT), R1 AND C1 FORM AN RC CIRCUIT. C1 GRADUALLY CHARGES TO +9 VOLTS THROUGH R1. WHEN THE VOLTAGE ON C1 EXCEEDS THE REFERENCE VOLTAGE SUPPLIED TO THE NON-INVERTING INPUT OF THE OP-AMP, ITS OUTPUT SWINGS FROM HIGH TO LOW AND THE LED GLOWS. THE TIME DELAY CAN BE CHANGED BY ALTERING THE VALUES OF R1 AND C1 OR THE SETTING OF R2. DISCHARGE C1 (USE PUSHBUTTON SWITCH) FOR NEW CYCLE.

☐ **IC TIMERS** — THE SIMPLE CIRCUIT ABOVE IS THE KEY INGREDIENT OF MOST IC TIMERS. MOST INCLUDE AN OUTPUT FLIP-FLOP TO GIVE DEFINITE HIGH OR LOW OUTPUT. SOME INCLUDE A BINARY COUNTER THAT ADVANCES ONE COUNT PER DELAY PERIOD (OR CYCLE). THE TIMER IS RECYCLED EACH TIME THE COUNT ADVANCES. A DECODER AT THE COUNTER OUTPUT ALLOWS TOTAL DELAYS OF FROM DAYS TO A YEAR OR MORE TO BE SELECTED. BOTH BIPOLAR AND CMOS TIMERS ARE AVAILABLE. ⇒ P.126

> **FAMOUS FACT:** ANALOG COMPUTERS USE OP-AMPS TO SOLVE COMPLEX EQUATIONS!

FUNCTION GENERATORS

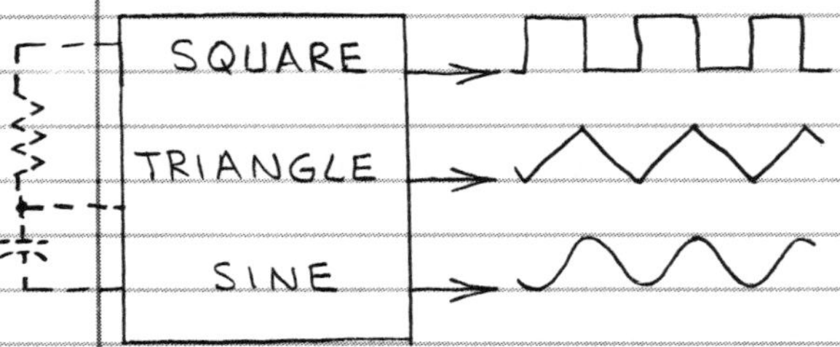

THESE IC'S GENERATE VARIOUS KINDS OF OUTPUT WAVES SUCH AS THOSE SHOWN HERE. THE FREQUENCY OF THE WAVES CAN BE CONTROLLED BY AN EXTERNAL RC CIRCUIT.

(SQUARE, TRIANGLE, SINE)

VOLTAGE REGULATORS

Voltage regulators convert a voltage applied to their input into a fixed or variable (but usually lower) voltage. In most a small, fixed reference voltage (usually a volt or so) is applied to the non-inverting input of an op-amp. The reference voltage (or V_{REF}) is then amplified by the ratio of the feedback and input resistors (the gain). If one of the resistors is a potentiometer, the output voltage (V_{OUT}) can be varied from V_{REF} to $+V$ (the chip supply voltage). Actual IC regulators include extra transistors to provide V_{REF} and to allow the chip to drive loads that require more power than an op-amp alone can deliver.

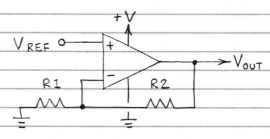

BASIC VOLTAGE REGULATOR

☐ IC REGULATORS — Many types of fixed and variable output IC regulators are available. Most are installed in packages made of metal or having metal tabs to help radiate excessive heat into the surrounding air. CAUTION: Manufacturer's operating instructions AND standard safety precautions must be followed for best results. [P.125 →]

OTHER LINEAR IC's

There are numerous special function linear IC's, many of which incorporate op-amps. For example:

☐ AUDIO AMPLIFIERS — Many kinds available. Some include TWO amplifiers on one chip (for stereo).

☐ PHASE-LOCKED LOOPS — Based on an old but clever idea in which an on-chip oscillator duplicates (or tracks) the frequency of an incoming signal. Used to detect the presence of certain frequencies (like Touch-Tone® tones) and to demodulate FM radio signals.

☐ OTHER LINEAR IC's — Included are many kinds of chips for telephone, radio, television and computer communications. Also, many kinds of IC's that detect temperature, light and pressure.

95

8. CIRCUIT ASSEMBLY TIPS

THERE ARE SEVERAL WAYS TO MAKE EITHER TEMPORARY OR PERMANENT VERSIONS OF ELECTRONIC CIRCUITS. IN THIS CHAPTER WE'LL LOOK AT SOME CIRCUIT ASSEMBLY TIPS YOU MAY FIND HELPFUL.

TEMPORARY CIRCUITS

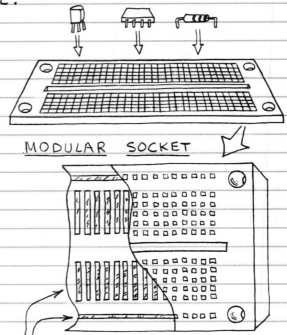

MODULAR SOCKET

CUTAWAY SHOWING COMMON TERMINAL CONNECTIONS.

IT'S ALWAYS WISE TO BUILD A TEMPORARY VERSION OF A CIRCUIT BEFORE ASSEMBLING IT IN PERMANENT FORM. YOU CAN THEN MAKE CHANGES AND FIND OUT HOW WELL THE CIRCUIT WORKS. BY FAR THE MOST IMPORTANT TOOL FOR TEMPORARY CIRCUIT ASSEMBLY IS THE PLASTIC SOLDERLESS MODULAR BREADBOARD SOCKET. IT'S A GOOD IDEA TO KEEP SEVERAL ON YOUR WORKBENCH. THEY WILL LET YOU BUILD ENTIRE CIRCUITS IN MINUTES. USE "JUMPER" WIRES TO INTERCONNECT PARTS WHOSE LEADS ARE NOT INSERTED IN THE SAME ROW OF TERMINALS. TO AVOID BENDING THEIR PINS (AND PRICKING YOUR FINGERS), INSTALL AND REMOVE IC'S CAREFULLY.

HINT: INSTALL SOCKET ON BASE AND ADD POTENTIOMETERS, BATTERY, LEDs, SWITCHES, ETC.

PERMANENT CIRCUITS

WITH THE EXCEPTION OF SOME VERY SIMPLE CIRCUITS, MOST PERMANENT CIRCUITS ARE ASSEMBLED ON SOME FORM OF CIRCUIT BOARD.

☐ PERFORATED BOARD CONSTRUCTION — COMPONENT LEADS ARE INSERTED THROUGH PERFORATIONS IN A PHENOLIC OR SIMILAR BOARD AND SOLDERED TOGETHER ON THE BACK SIDE OF THE BOARD. OFTEN INSULATED CONNECTION WIRES MUST BE USED. ONCE ASSEMBLED, "PERFBOARD" CIRCUITS ARE DIFFICULT TO REPAIR SINCE COMPONENT LEADS ARE OFTEN TWISTED AND SOLDERED.

COMPONENT CONNECTION (REQUIRES SOLDER)

☐ **WIRE-WRAP** — FASTEST WAY TO ASSEMBLE CIRCUITS THAT USE MORE THAN A FEW IC'S. USE WIRE-WRAP IC SOCKETS (WITH SQUARE CONNECTION PINS). BOTH HAND AND MOTOR POWERED WRAPPING TOOLS ARE AVAILABLE. IF YOU USE THE KIND THAT REQUIRES SOME OF THE WIRE'S INSULATION BE REMOVED, WRAP A FEW TURNS OF INSULATED WIRE AROUND THE CONNECTION PIN TO STRENGTHEN THE CONNECTION.

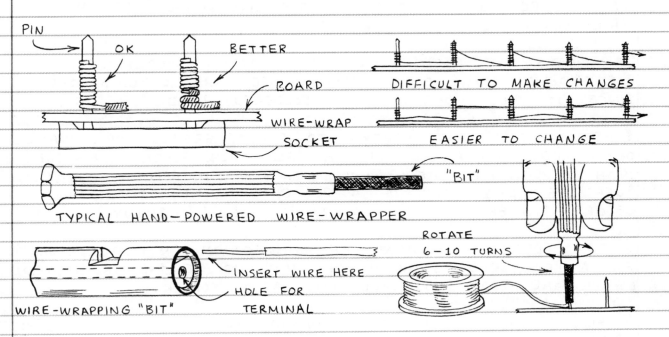

☐ **PRINTED CIRCUIT (PC)** — PROVIDES THE NEATEST AND MOST PROFESSIONAL APPEARING COMPLETED CIRCUIT. SOCKETS NOT REQUIRED, BUT COMPONENT LEADS MUST BE SOLDERED TO THE COPPER PATTERNS ON THE BOARD. THERE ARE MANY TYPES OF PC BOARDS. TWO TYPES USED BY EXPERIMENTERS ARE:

1. PRE-ETCHED PERFORATED GRID BOARDS HAVE A ROUND, COPPER FOIL SOLDER PAD AT EACH HOLE. ON MANY BOARDS ROWS OF HOLES ARE CONNECTED BY COMMON COPPER FOIL STRIPS (LIKE A SOLDERLESS BREADBOARD). IT'S USUALLY NECESSARY TO JOIN SOME OF THE CONTACTS ON THE BOARD WITH "JUMPERS" (SHORT LENGTHS OF INSULATED HOOKUP OR WRAPPING WIRE).

2. CUSTOM PC BOARDS ARE MADE BY APPLYING A TAPE OR CHEMICAL COATING (THE "RESIST") TO THE CLEAN COPPER FOIL OF A PC BOARD. THE UNCOATED COPPER IS THEN CHEMICALLY ETCHED AWAY, LEAVING BEHIND A FOIL WIRING PATTERN. HOLES ARE DRILLED FOR COMPONENT LEADS. TAKES LOTS OF TIME, BUT PRODUCES NEAT CIRCUITS.

HOW TO SOLDER

Good soldering practices are essential for reliable operation of a circuit with soldered connections. Here are six steps for successful soldering:

1. Always use a low-wattage soldering iron (25 to 40 watts). Be sure to tin the tip according to the manufacturer's instructions.

2. Always use <u>rosin core</u> solder when soldering electronic components. <u>Never use acid core solder since it will corrode the soldered lead.</u>

3. Solder does <u>not</u> adhere to paint, grease, oil, wax or melted insulation. Remove all such foreign matter with a solvent, steel wool or fine sandpaper. <u>Always</u> buff the copper foil of a PC board with steel wool before soldering. (The copper should be shiny.)

4. To solder, first heat the connection (<u>not</u> the solder!) for a few seconds with the hot tip of the iron. Then leave the iron in place and apply solder.

5. Allow the solder to flow through and around the connection before removing the iron. Don't apply too much solder or move the connection before it cools.

6. Keep the iron's tip clean and shiny. Wipe off debris with a damp sponge or cloth.

SOLDERING PRECAUTIONS

1. A hot soldering iron can burn a finger or even start a fire. Use care!

2. Unplug the iron when you're not using it.

3. Be sure the power cord is not where you can trip over it.

(Hold parts in place with tape)

POWERING ELECTRONIC CIRCUITS

☐ BATTERY POWER — MANY CIRCUITS USE SO LITTLE POWER THEY CAN BE POWERED BY BATTERIES. THIS KEEPS THE COMPLETED CIRCUIT COMPACT AND ALLOWS IT TO BE OPERATED ANYWHERE.

☐ SOLAR POWER — SOLAR CELLS CAN POWER YOUR CIRCUITS DIRECTLY. OR YOU CAN USE AN ARRAY OF SOLAR CELLS TO CHARGE A RECHARGEABLE BATTERY.

☐ LINE POWER — THE SIMPLEST LINE POWERED SUPPLY IS THE SO-CALLED AC ADAPTER. THESE MODULAR UNITS ARE COMPACT AND EASY TO USE. UNITS HAVING VARIOUS OUTPUT VOLTAGES AND CURRENTS ARE AVAILABLE. YOU CAN MAKE YOUR OWN LINE POWERED SUPPLY USING AN IC VOLTAGE REGULATOR.

☐ <u>CAUTION</u> — SAFETY SHOULD BE YOUR FIRST CONCERN WHEN BUILDING YOUR OWN LINE POWERED SUPPLY. THE POWER CORD <u>MUST</u> BE PROTECTED FROM THE SHARP EDGES OF A HOLE DRILLED IN A METAL CABINET. (USE A PLASTIC STRAIN RELIEF.) <u>ALL</u> CONNECTIONS TO THE AC LINE <u>MUST</u> BE INSIDE A FULLY ENCLOSED HOUSING! LEAVING SUCH CONNECTIONS EXPOSED IS A POTENTIAL SHOCK HAZARD. MAKE SURE <u>ALL</u> COMPONENTS THAT ARE CONNECTED TO THE AC LINE (SWITCHES, FUSES, TRANSFORMERS, ETC.) MEET OR EXCEED THE POWER REQUIREMENT OF YOUR CIRCUIT.

SUMMING UP CIRCUIT ASSEMBLY

THE REMAINDER OF THIS BOOK INCLUDES MANY CIRCUITS YOU CAN QUICKLY ASSEMBLE ON A SOLDERLESS BREADBOARD. CHANCES ARE YOU'LL WANT TO MAKE PERMANENT VERSIONS OF SOME. FOR BEST RESULTS, PLAN THE PROJECT CAREFULLY. A NEATLY ASSEMBLED PROJECT WILL BE MORE RELIABLE THAN ONE HASTILY ASSEMBLED.

SLOPPY PROJECT

NEAT PROJECT

9. 100 ELECTRONIC CIRCUITS

Here's a collection of 100 electronic circuits. I've assembled each circuit to make sure all of them work.

☐ SELECTING AND SUBSTITUTING COMPONENTS — You can find most of the components at Radio Shack stores. Save time and make a list of what you need before you visit Radio Shack. (You can find current catalog numbers in the latest Radio Shack catalog.) If a component is unavailable, try elsewhere. Sometimes you can substitute components. For example, it's often OK to substitute NPN switching transistors for one another (2N3904 for 2N2222, etc.). Nearby values of resistors and capacitors can often be used (1.2K for 1K resistor, 0.33μF for 0.47μF capacitor, etc.). Always follow appropriate voltage and power ratings!

☐ WHEN A CIRCUIT DOESN'T WORK — Make sure the circuit is receiving adequate power. If it is or if you smell or feel a hot component, immediately disconnect the power and follow these steps: ① Recheck ALL connections. (Is a wire missing? Is an IC pin bent? Is a solder connection bad? Is a wire "shorted"? Is a diode backwards?) ② Is a component defective? ③ Sometimes, especially when power supply leads are more than six inches long, IC circuits will work improperly or not at all unless you connect a 0.1 μF capacitor across the power supply pins of each chip. It may also be necessary to connect a 1 to 10 μF capacitor across the power leads where they enter the board. ④ Does the published circuit contain an error?

☐ SAFETY FIRST — Be sure to follow appropriate precautions when working with AC line powered circuits. Be careful when soldering. Circuits with speakers can produce very loud sounds. Keep your distance, and don't use headphones.

☐ GOING FURTHER — Try experimenting with the values of components in RC circuits (p. 37). Try substituting other output devices in circuits that drive a relay, piezo buzzer, etc. (Be sure to follow voltage and current ratings. Use Ohm's law and, if necessary, add a series resistor to reduce current.) Before building a permanent version of a circuit, always assemble and test a breadboard version. Finally, be sure to buy Radio Shack's current "Semiconductor Reference Guide" and "Engineer's Notebook." For more advanced circuits and information about new developments, read my column ("The Electronics Scientist") in Computers & Electronics.

DIODE CIRCUITS

THE VARIOUS KINDS OF DIODES HAVE <u>MANY</u> APPLICATIONS. HERE ARE SOME TYPICAL CIRCUITS:

SMALL SIGNAL DIODES AND RECTIFIERS

☐ VOLTAGE REGULATOR   ☐ VOLTAGE DROPPER

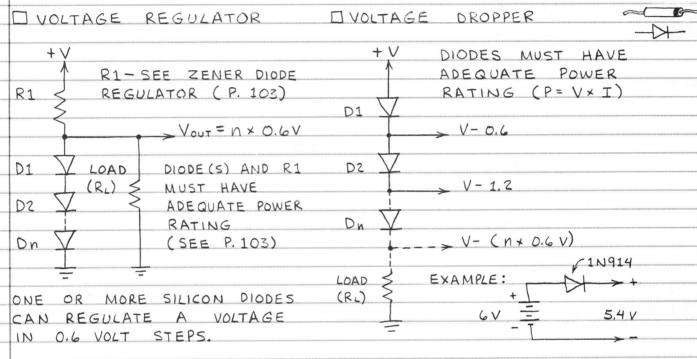

ONE OR MORE SILICON DIODES CAN REGULATE A VOLTAGE IN 0.6 VOLT STEPS.

R1 — SEE ZENER DIODE REGULATOR (P. 103)

$V_{out} = n \times 0.6V$

DIODE(S) AND R1 MUST HAVE ADEQUATE POWER RATING (SEE P. 103)

DIODES MUST HAVE ADEQUATE POWER RATING ($P = V \times I$)

EXAMPLE: 6V → 5.4V (1N914)

☐ 9 VOLT POWER SUPPLY

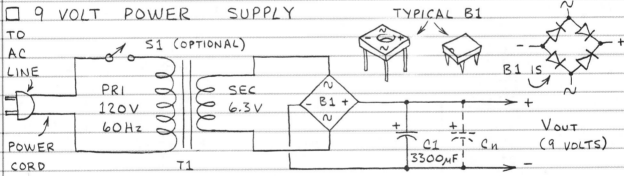

TYPICAL B1

B1 IS ⟨bridge⟩

THIS IS A BASIC AC LINE OPERATED 9 VOLT POWER SUPPLY. FOR LOW RIPPLE (SUPERIMPOSED AC AT V_{out}), USE LARGE VALUE FOR C1. (OK TO ADD ONE OR MORE CAPACITORS (Cn) IN PARALLEL WITH C1 FOR MORE CAPACITANCE.) CAPACITORS MUST HAVE A DC WORKING VOLTAGE (WVDC) OF AT LEAST 12 VOLTS. RECTIFIER BRIDGE B1 MUST HAVE PEAK INVERSE VOLTAGE (PIV) OF AT LEAST 12 VOLTS. T1 AND B1 MUST HAVE ADEQUATE POWER AND CURRENT RATINGS. (USE OHMS LAW...)
<u>CAUTION</u>: YOU <u>MUST</u> INSULATE OR ENCLOSE ALL EXPOSED AC LINE CONNECTIONS! THE POWER CORD <u>MUST</u> BE UNPLUGGED WHEN YOU ASSEMBLE OR SERVICE THE CIRCUIT!

☐ VOLTAGE DOUBLERS

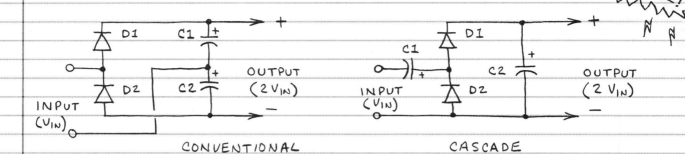

CONVENTIONAL   CASCADE

THESE CIRCUITS APPROXIMATELY DOUBLE AN INCOMING AC VOLTAGE. THE OUTPUT IS DC. USE CAPACITORS AND DIODES RATED FOR TWICE THE INPUT VOLTAGE. OUTPUT RIPPLE (∿) CAN BE REDUCED BY USING LARGE VALUES FOR C1 AND C2.

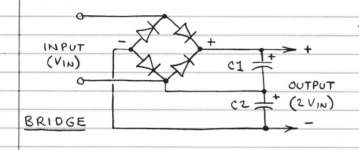

THE BRIDGE DOUBLER IS MORE EFFICIENT THAN THE CONVENTIONAL AND CASCADE DOUBLERS. SINCE 4-DIODE BRIDGE RECTIFIERS ARE AVAILABLE, IT'S EASY TO MAKE.

☐ VOLTAGE TRIPLER   ☐ VOLTAGE QUADRUPLER

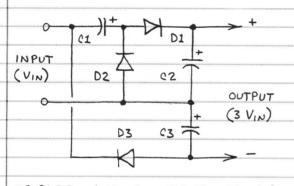

TRIPLES AND CONVERTS TO DC AN INCOMING AC VOLTAGE. C2, D1, D2 AND D3 RATED AT $>2 V_{IN}$.

QUADRUPLES AND CONVERTS TO DC AN INCOMING AC VOLTAGE. ALL COMPONENTS RATED AT $>2 V_{IN}$.

☐ CASCADE MULTIPLIER

ADD MORE STAGES FOR MORE MULTIPLICATION. ALL COMPONENTS RATED AT $>2V_{IN}$.

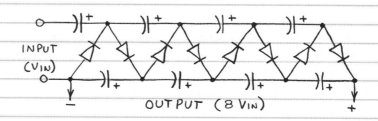

CAUTION: THESE CIRCUITS CAN PRODUCE HIGH VOLTAGE!

ZENER DIODE CIRCUITS

☐ VOLTAGE REGULATOR

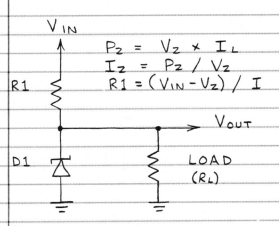

$P_z = V_z \times I_L$
$I_z = P_z / V_z$
$R1 = (V_{IN} - V_z) / I$

I_L = MAXIMUM LOAD CURRENT
I_z = MAXIMUM ZENER CURRENT
I = CURRENT THROUGH R1
V_z = ZENER DIODE VOLTAGE
P_z = ZENER DIODE POWER

THIS CIRCUIT SUPPLIES A STEADY VOLTAGE (V_{OUT}) TO A LOAD FROM AN UNREGULATED SUPPLY (LIKE A BATTERY). V_{IN} CAN VARY BUT MUST BE AT LEAST 1 VOLT ABOVE DESIRED V_{OUT}. I_L CAN VARY FROM 0 MA TO PLANNED MAXIMUM VALUE. I DOES NOT CHANGE IF I_L FALLS TO 0 MA. SINCE $I = I_L + I_z$, I_z RISES AS I_L FALLS. IN OTHER WORDS, THE REGULATOR ALWAYS USES THE SAME CURRENT, EVEN WHEN THE LOAD IS REMOVED. CAUTION: D1 AND R1 MUST HAVE PROPER POWER RATING. USE OHM'S LAW (P.14).

EXAMPLE: A RADIO DRAWS FROM 20 TO 50 MA FROM A 9 VOLT BATTERY. TO POWER IT FROM A 12 VOLT BATTERY, USE A 9 VOLT, 1/2 WATT ZENER DIODE. R1 SHOULD BE CLOSE TO 60 Ω AND RATED FOR AT LEAST 0.15 W.

EXAMPLE CIRCUIT:

12 V (V_{IN}), R1 = 60 Ω, 1/2 W, D1 = 9 V, 1/2 W, 9 V (V_{OUT})

☐ WAVEFORM CLIPPER

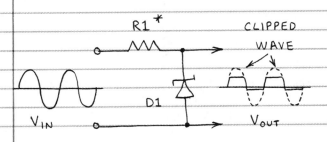

THIS CIRCUIT IS HANDY FOR REDUCING AN INCOMING SIGNAL VOLTAGE TO A LOWER, MORE MANAGEABLE LEVEL. IT CAN ALSO CONVERT A SINE OR TRIANGLE WAVE TO AN APPROXIMATION OF A SQUARE WAVE. *R1: SEE ABOVE (LET I = 2 MA MINIMUM).

☐ DUAL WAVEFORM CLIPPER

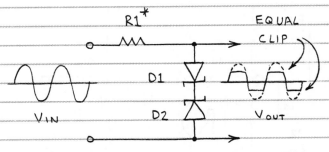

THIS IS A SYMMETRICAL FORM OF THE ADJACENT CIRCUIT. IT CLIPS BOTH HALVES OF AN INCOMING WAVE EQUALLY (IF V_z = D1 = D2). USE TO PROTECT SPEAKERS AND PHONES FROM EXCESSIVE SIGNAL LEVELS OR TO MAKE SQUARE WAVES.

TRANSISTOR CIRCUITS

INTEGRATED CIRCUITS RECEIVE MORE ATTENTION, BUT BOTH BIPOLAR AND FIELD-EFFECT TRANSISTORS HAVE MANY APPLICATIONS.

BIPOLAR TRANSISTOR CIRCUITS

☐ MOISTURE METER   ☐ MOISTURE ACTUATED RELAY

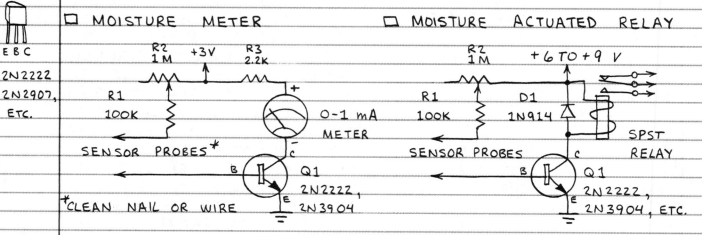

THIS CIRCUIT WILL MEASURE THE MOISTURE LEVEL OF SOIL IN YOUR GARDEN. ADJUST R2 FOR METER READING OF 1 mA WHEN SOIL MOISTURE IS AT DESIRED LEVEL. METER WILL THEN INDICATE LOWER MOISTURE LEVELS.

THIS CIRCUIT CLOSES A RELAY (6-9 V, 500 OHM COIL) WHEN MOISTURE LEVEL EXCEEDS A LEVEL SET BY R2. RELAY CONTACTS THEN SWITCH ON A LIGHT OR OTHER DEVICE. CAN DETECT RAIN.

☐ METRONOME   ☐ LIGHT FLASHER

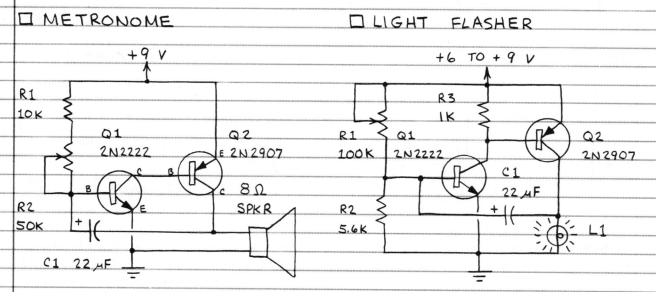

A METRONOME MARKS TIME BY PRODUCING A REGULAR SEQUENCE OF "CLICKS" OR "POCKS." ADJUST THE CLICK RATE BY ADJUSTING R2 OR CHANGING C1'S VALUE.

THIS CIRCUIT PRODUCES BRIGHT FLASHES EVERY SECOND OR SO. R1 CONTROLS FLASH RATE. USE NO. 122 OR 222 MINIATURE LAMP FOR L1.

☐ SIREN

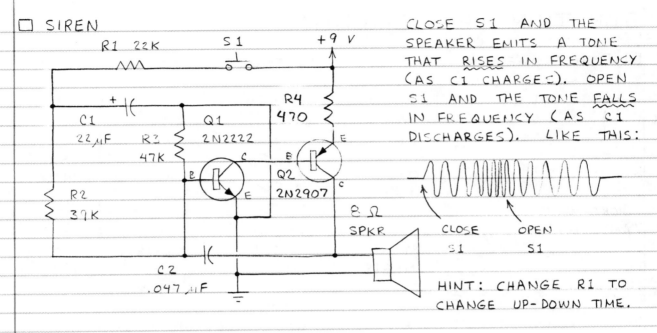

CLOSE S1 AND THE SPEAKER EMITS A TONE THAT RISES IN FREQUENCY (AS C1 CHARGES). OPEN S1 AND THE TONE FALLS IN FREQUENCY (AS C1 DISCHARGES). LIKE THIS:

HINT: CHANGE R1 TO CHANGE UP-DOWN TIME.

☐ HIGH VOLTAGE POWER SUPPLY

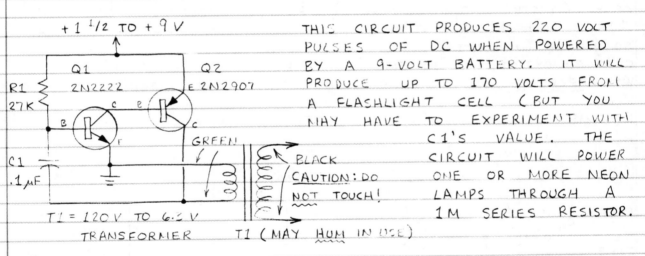

THIS CIRCUIT PRODUCES 220 VOLT PULSES OF DC WHEN POWERED BY A 9-VOLT BATTERY. IT WILL PRODUCE UP TO 170 VOLTS FROM A FLASHLIGHT CELL (BUT YOU MAY HAVE TO EXPERIMENT WITH C1'S VALUE. THE CIRCUIT WILL POWER ONE OR MORE NEON LAMPS THROUGH A 1M SERIES RESISTOR.

☐ BURGLAR ALARM

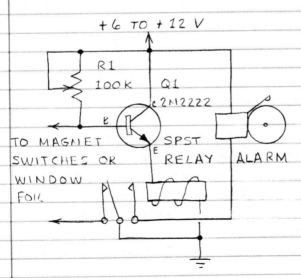

ALARM SOUNDS AND STAYS ON (UNTIL POWER IS DISCONNECTED) WHEN MAGNET SWITCH IS OPENED OR WINDOW FOIL IS BROKEN. R1: SET TO MAXIMUM VALUE. DISCONNECT ONE LEAD TO SWITCHES/FOIL. THEN REDUCE VALUE OF R1 TO JUST PAST POINT WHERE ALARM SOUNDS. CIRCUIT USES ONLY 0.3 MILLIAMPERES AT 6 VOLTS. RELAY: USE 6V, 500Ω UNIT WHEN SUPPLY IS 6-9 V. USE 12 V, 1200Ω UNIT WHEN SUPPLY IS 12 V. NOTE: ASSEMBLE, INSTALL AND CONCEAL THIS SYSTEM WITH CARE.

JUNCTION FET CIRCUITS

☐ ELECTROMETER

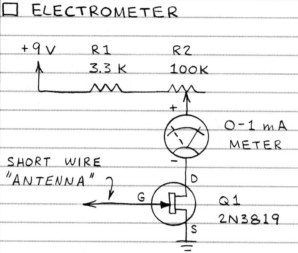

THIS CIRCUIT DETECTS STATIC ELECTRICITY FROM A CHARGED OBJECT (PLASTIC COMB, ETC.) OVER A FOOT AWAY! ADJUST R1 SO METER INDICATES 1 MA. CHARGED OBJECT NEAR "ANTENNA" WILL <u>DECREASE</u> METER READING.

☐ TOUCH SWITCH

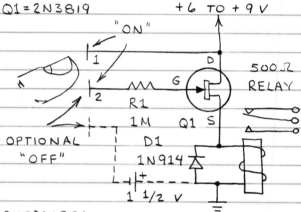

OUTDOORS:
AWAY FROM POWER LINES, BRIEFLY TOUCH "ON" CONTACTS TO ACTUATE RELAY. TOUCH CONTACT 2 ONLY TO DEACTUATE RELAY.
INDOORS:
MAY BE NECESSARY TO INCLUDE OPTIONAL "OFF" CIRCUIT.

☐ TIMER

DELAY TIME: UP TO 1.5 MINUTES

SET S1 TO "RESET" (BUZZER WILL SOUND). THEN SET S1 TO "TIME." BUZZER WILL BE SILENT UNTIL DELAY IS COMPLETE. BUZZER WILL THEN SOUND. INCREASE C1 OR R1 FOR LONGER DELAYS. REDUCE R2'S RESISTANCE DURING RESET MODE (SPEEDS UP RESET).

☐ AUDIO MIXER

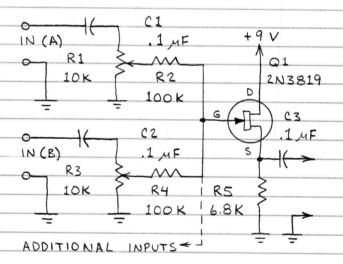

THIS CIRCUIT ALLOWS TWO (OR MORE) MICROPHONES OR OTHER DEVICES TO BE CONNECTED TO THE SAME AMPLIFIER. R1 AND R3 CONTROL ATTENUATION OF EACH INPUT. THEREFORE R1 AND R3 ARE <u>BALANCE CONTROLS</u>.

POWER MOSFET (DMOS, VMOS, ETC.) CIRCUITS

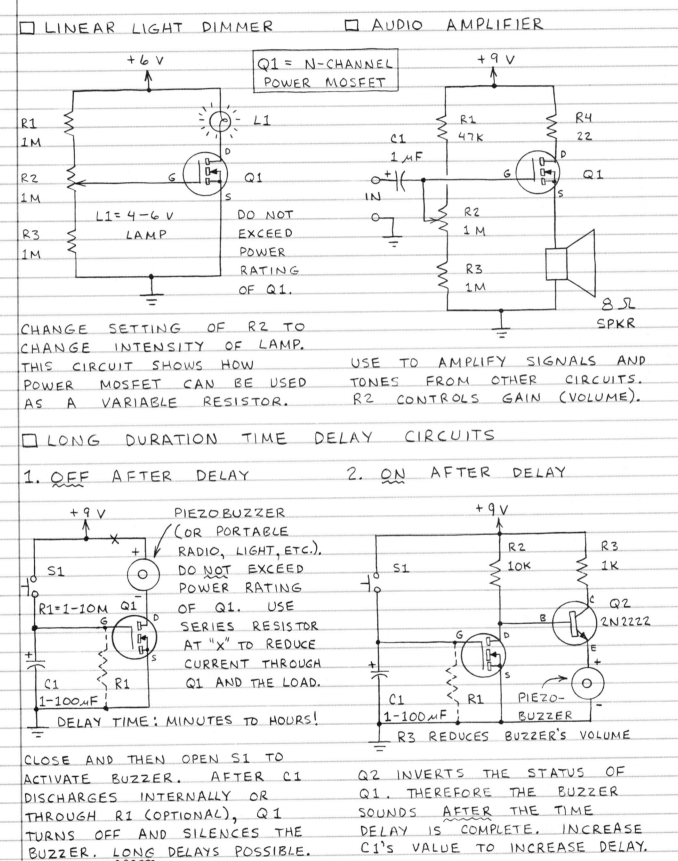

☐ LINEAR LIGHT DIMMER   ☐ AUDIO AMPLIFIER

Q1 = N-CHANNEL POWER MOSFET

R1 1M, R2 1M, R3 1M, L1 = 4-6V LAMP, +6V

DO NOT EXCEED POWER RATING OF Q1.

CHANGE SETTING OF R2 TO CHANGE INTENSITY OF LAMP. THIS CIRCUIT SHOWS HOW POWER MOSFET CAN BE USED AS A VARIABLE RESISTOR.

C1 1μF, R1 47K, R2 1M, R3 1M, R4 22, +9V, 8Ω SPKR

USE TO AMPLIFY SIGNALS AND TONES FROM OTHER CIRCUITS. R2 CONTROLS GAIN (VOLUME).

☐ LONG DURATION TIME DELAY CIRCUITS

1. OFF AFTER DELAY   2. ON AFTER DELAY

+9V, S1, R1 = 1-10M, C1 = 1-100μF, Q1

PIEZOBUZZER (OR PORTABLE RADIO, LIGHT, ETC.). DO NOT EXCEED POWER RATING OF Q1. USE SERIES RESISTOR AT "X" TO REDUCE CURRENT THROUGH Q1 AND THE LOAD.

DELAY TIME: MINUTES TO HOURS!

CLOSE AND THEN OPEN S1 TO ACTIVATE BUZZER. AFTER C1 DISCHARGES INTERNALLY OR THROUGH R1 (OPTIONAL), Q1 TURNS OFF AND SILENCES THE BUZZER. LONG DELAYS POSSIBLE.

+9V, S1, R2 10K, R3 1K, Q2 2N2222, C1 1-100μF, R1, PIEZOBUZZER

R3 REDUCES BUZZER'S VOLUME

Q2 INVERTS THE STATUS OF Q1. THEREFORE THE BUZZER SOUNDS AFTER THE TIME DELAY IS COMPLETE. INCREASE C1'S VALUE TO INCREASE DELAY.

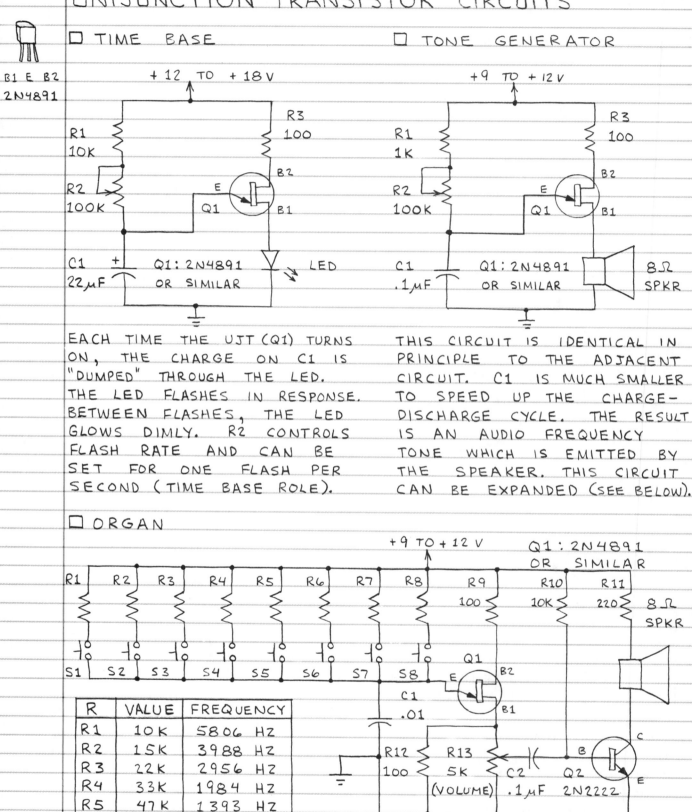

☐ RAMP GENERATOR

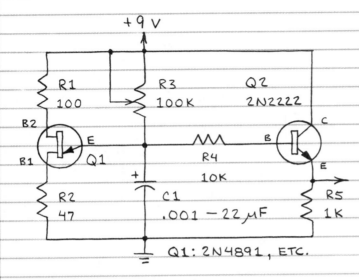

Q1: 2N4891, ETC.

Q2 SAMPLES THE VOLTAGE ON C1 AND OUTPUTS IT AS A RAMP (OR "SAWTOOTH" WAVE). R3 CONTROLS THE RATE AT WHICH RAMPS ARE PRODUCED.

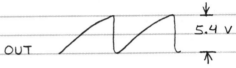

RAMPS SUPPLY GRADUALLY INCREASING VOLTAGE TO MANY KINDS OF CIRCUITS.

☐ CHIRP GENERATOR

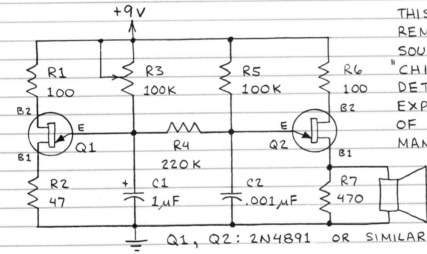

Q1, Q2: 2N4891 OR SIMILAR

PIEZO TWEETER SPEAKER

THIS CIRCUIT PRODUCES A REMARKABLE VARIETY OF SOUNDS. AS SHOWN, IT "CHIRPS" AT A RATE DETERMINED BY R3. EXPERIMENT WITH VALUES OF C1, R5 AND C2 FOR MANY OTHER EFFECTS.

☐ VOLTAGE SENSITIVE OSCILLATOR

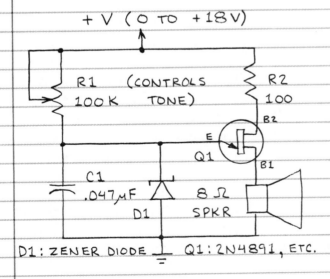

D1: ZENER DIODE   Q1: 2N4891, ETC.

THIS CIRCUIT SOUNDS A TONE WHEN V_{IN} IS BELOW V_Z OF D1. SELECT V_Z OF D1 FOR DESIRED TURN-OFF LEVEL. THIS CIRCUIT CAN BE USED TO INDICATE WHEN VOLTAGE OF A BATTERY (WHICH CAN BE POWERING ANOTHER CIRCUIT) FALLS BELOW A CERTAIN LEVEL. THIS IS AN EXCELLENT EXAMPLE OF A SIMPLE, YET SOPHISTICATED, CIRCUIT.

THYRISTOR CIRCUITS

SILICON CONTROLLED RECTIFIERS AND TRIACS HAVE MANY APPLICATIONS AS SOLID-STATE SWITCHES.

SCR CIRCUITS

☐ LATCHING SWITCH   ☐ TEST CIRCUIT

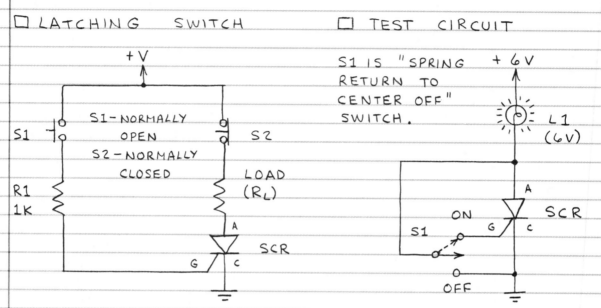

CLOSE S1 TO TURN ON THE SCR AND SUPPLY CURRENT TO THE LOAD. THE SCR WILL REMAIN ON AFTER S1 IS OPENED UNLESS THE LOAD IS A DC MOTOR OR UNTIL S2 IS BRIEFLY OPENED.

S1 IS AN SPDT SWITCH. IN THE "ON" POSITION, THE SCR IS TURNED ON AND THE LAMP GLOWS. IN THE "OFF" POSITION, CURRENT IS SHUNTED AWAY FROM SCR, THUS TURNING IT OFF.

☐ CAPACITOR DISCHARGE LED FLASHER

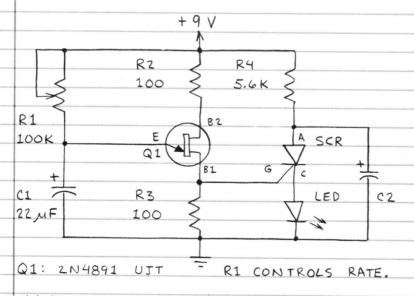

Q1: 2N4891 UJT   R1 CONTROLS RATE.

WHEN THE SCR IS OFF, C2 CHARGES THROUGH R4. WHEN THE SCR IS TURNED ON BY A PULSE FROM UJT Q1, THE CHARGE IN C1 IS RAPIDLY "DUMPED" THROUGH THE LED. THE SCR (AND LED) THEN TURNS OFF SINCE THERE IS NO LONGER SUFFICIENT HOLDING CURRENT. THE CYCLE THEN REPEATS.

TRIAC CIRCUITS

☐ TEST CIRCUIT

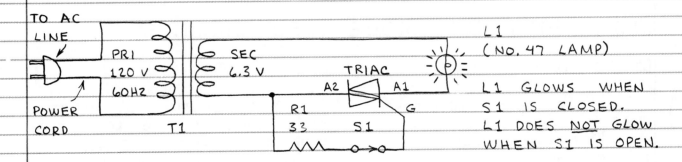

L1 GLOWS WHEN S1 IS CLOSED. L1 DOES NOT GLOW WHEN S1 IS OPEN.

CAUTION:
TRIACS ARE DESIGNED SPECIFICALLY FOR AC OPERATION. BE SURE TO OBSERVE COMMON SENSE SAFETY WHEN WORKING WITH HOUSEHOLD LINE CURRENT! MAKE SURE ALL CONNECTIONS TO AC LINE ARE INSULATED OR ENCLOSED.

☐ LIGHT DIMMER CIRCUITS

1. 6.3 VOLT DIMMER

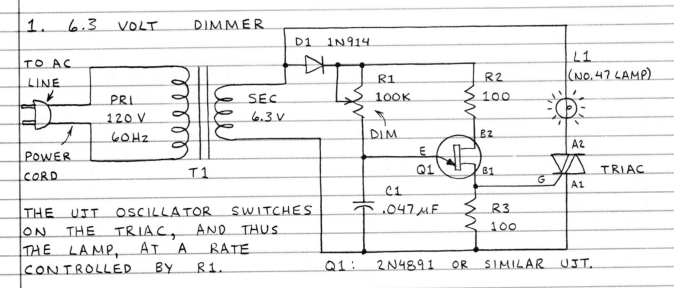

THE UJT OSCILLATOR SWITCHES ON THE TRIAC, AND THUS THE LAMP, AT A RATE CONTROLLED BY R1.

Q1: 2N4891 OR SIMILAR UJT.

2. 120 VOLT DIMMER

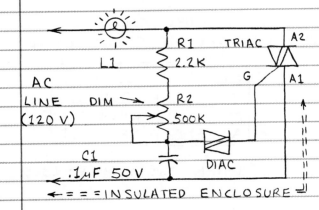

THIS IS HOW MANY HOUSEHOLD DIMMER SWITCHES ARE MADE. L1 CAN BE UP TO 100 WATT (120 VOLT) LAMP. USE HEAT SINK ON TRIAC IF IT BECOMES HOT. THE DIAC IS A BIDIRECTIONAL TRIGGER DIODE.
CAUTION: THIS CIRCUIT MUST BE FULLY ENCLOSED AT ALL TIMES WHEN POWER IS APPLIED!

PHOTONIC CIRCUITS

CIRCUITS USING PHOTONIC COMPONENTS ARE AMONG THE MOST VERSATILE AND INTERESTING OF ALL CIRCUITS.

LIGHT EMITTING DIODE (LED) CIRCUITS

☐ LED DRIVE CIRCUIT

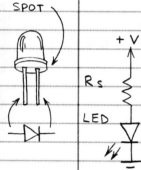
FLAT SPOT

A SERIES RESISTOR MUST BE USED TO LIMIT CURRENT THROUGH AN LED. (EXCEPTIONS: CERTAIN PULSE CIRCUITS AND IC LED DRIVERS.)

$$R_S = \frac{+V - (V_{LED})}{I_{LED}}$$

EXAMPLE: ASSUME YOU WANT TO OPERATE A RED LED AT A FORWARD CURRENT (I_{LED}) OF 10 MA (OR 0.01 AMPERE) FROM A 5 VOLT SUPPLY. THE DATA SHEET FOR THE LED LISTS AN LED VOLTAGE (V_{LED}) OF 1.7 VOLTS. THEREFORE R_S IS $(5-1.7)/.01$ OR 330 OHMS.

☐ VARIABLE BRIGHTNESS LED

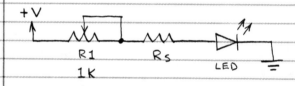

ADJUST R1 TO ALTER CURRENT THROUGH THE LED, THUS VARYING ITS BRIGHTNESS. R_S MUST BE USED (SEE ABOVE).

☐ POLARITY INDICATOR

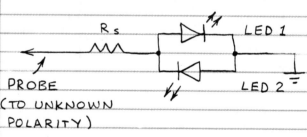

THIS CIRCUIT INDICATES THE POLARITY OF A VOLTAGE. R_S MUST BE USED (SEE ABOVE).

LED	+	−	AC (±)
1	ON	OFF	ON
2	OFF	ON	ON

☐ TRI-STATE POLARITY INDICATOR

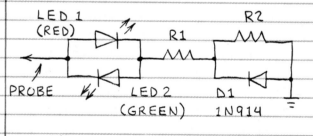

THIS IS A MORE COLORFUL VERSION OF THE CIRCUIT ABOVE.

V_{IN}	COLOR
+	RED
−	GREEN
AC (±)	YELLOW*

$$R1 = \frac{V_{IN} - (V_{LED2} + 0.6)}{I_{LED2}} \qquad R1 + R2 = \frac{V_{IN} - V_{LED1}}{I_{LED1}}$$

*WHEN TWO-CHIP LED USED.

☐ DUAL LED FLASHER

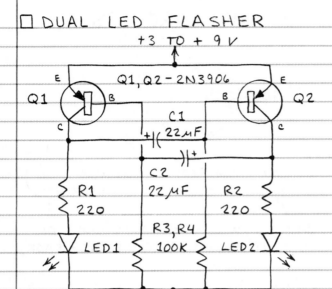

THIS CIRCUIT IS CALLED A FREE-RUNNING MULTIVIBRATOR. IT'S IDENTICAL TO A FLIP-FLOP THAT TRIGGERS ITSELF REPEATEDLY. Q1 AND Q2 ARE GENERAL PURPOSE PNP TRANSISTORS (2N3906, 2N2907, ETC.). R1 AND R2 LIMIT THE CURRENT TO THE LEDs (WHICH FLASH ALTERNATELY). INCREASING THE VALUES OF C1 AND C2 WILL SLOW THE FLASH RATE.

☐ VOLTAGE LEVEL INDICATOR

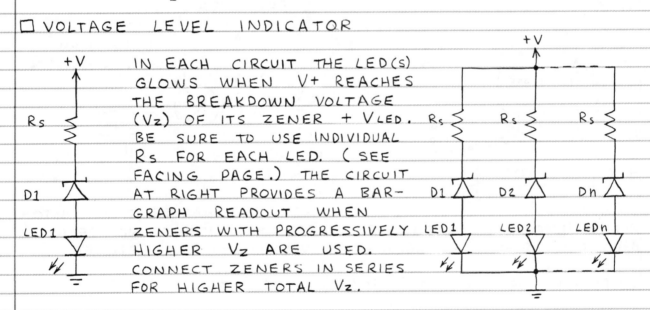

IN EACH CIRCUIT THE LED(s) GLOWS WHEN V+ REACHES THE BREAKDOWN VOLTAGE (V_z) OF ITS ZENER + V_{LED}. BE SURE TO USE INDIVIDUAL R_s FOR EACH LED. (SEE FACING PAGE.) THE CIRCUIT AT RIGHT PROVIDES A BAR-GRAPH READOUT WHEN ZENERS WITH PROGRESSIVELY HIGHER V_z ARE USED. CONNECT ZENERS IN SERIES FOR HIGHER TOTAL V_z.

☐ FLASHER LED + RELAY

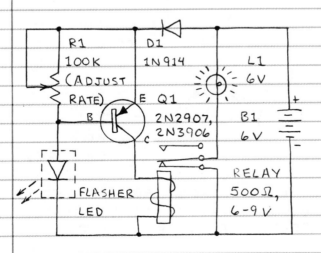

THE FLASHER LED INCLUDES A BUILT-IN IC THAT FLASHES THE LED SEVERAL TIMES EACH SECOND. THIS CIRCUIT SHOWS HOW TO "TAP" THIS FLASH RATE (VIA Q1) TO FORM AN ULTRA-SIMPLE PULSE GENERATOR THAT ACTUATES A RELAY AND, IN TURN, A LAMP. D1 IS NEEDED TO KEEP THE VOLTAGE TO THE FLASHER LED NEAR 5 VOLTS.

SEMICONDUCTOR LIGHT DETECTOR CIRCUITS

□ LIGHT METER CIRCUITS

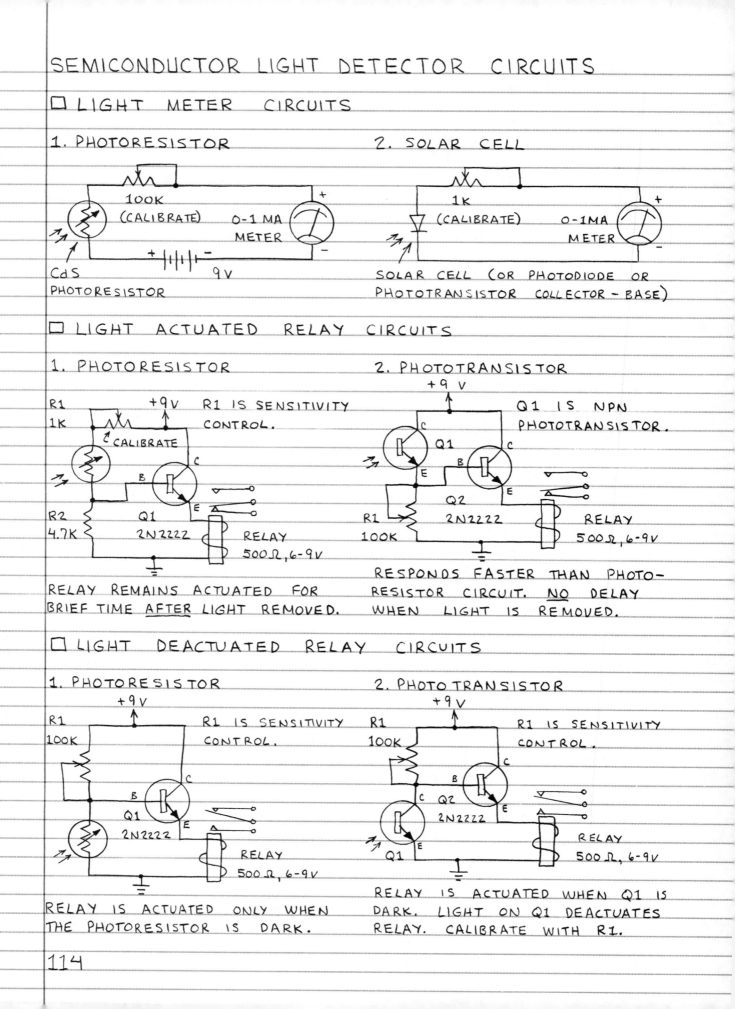

1. PHOTORESISTOR

100K (CALIBRATE), 0-1 MA METER, 9V, CdS PHOTORESISTOR

2. SOLAR CELL

1K (CALIBRATE), 0-1 MA METER

SOLAR CELL (OR PHOTODIODE OR PHOTOTRANSISTOR COLLECTOR-BASE)

□ LIGHT ACTUATED RELAY CIRCUITS

1. PHOTORESISTOR

R1 1K, +9V, CALIBRATE, R2 4.7K, Q1 2N2222, RELAY 500Ω, 6-9V

R1 IS SENSITIVITY CONTROL.

RELAY REMAINS ACTUATED FOR BRIEF TIME AFTER LIGHT REMOVED.

2. PHOTOTRANSISTOR

+9V, Q1, Q2 2N2222, R1 100Ω, RELAY 500Ω, 6-9V

Q1 IS NPN PHOTOTRANSISTOR.

RESPONDS FASTER THAN PHOTORESISTOR CIRCUIT. NO DELAY WHEN LIGHT IS REMOVED.

□ LIGHT DEACTUATED RELAY CIRCUITS

1. PHOTORESISTOR

+9V, R1 100K, Q1 2N2222, RELAY 500Ω, 6-9V

R1 IS SENSITIVITY CONTROL.

RELAY IS ACTUATED ONLY WHEN THE PHOTORESISTOR IS DARK.

2. PHOTOTRANSISTOR

+9V, R1 100K, Q2 2N2222, Q1, RELAY 500Ω, 6-9V

R1 IS SENSITIVITY CONTROL.

RELAY IS ACTUATED WHEN Q1 IS DARK. LIGHT ON Q1 DEACTUATES RELAY. CALIBRATE WITH R1.

114

☐ AUDIBLE LIGHT PROBE   ☐ SOLAR CELL BATTERY CHARGER

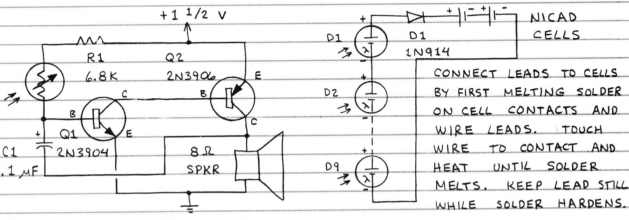

THIS IS AMONG THE MOST ENTERTAINING CIRCUITS IN THIS BOOK. VARIOUS PNP AND NPN TRANSISTORS CAN BE USED FOR Q1 AND Q2. THE TONE FROM THE SPEAKER INCREASES IN FREQUENCY AS THE INTENSITY OF LIGHT ON THE PHOTORESISTOR INCREASES. VERY SENSITIVE! TRY THIS: OPERATE CIRCUIT IN A DARK ROOM UNTIL TONE SLOWS TO A SERIES OF CLICKS. THEN SHINE BEAM FROM FLASHLIGHT ON THE PHOTORESISTOR...

USE NINE CELLS TO CHARGE TWO NICADS. CURRENT FROM SOLAR CELLS MUST NOT EXCEED MAXIMUM CHARGING RATE FOR NICADS! YOU CAN MONITOR CURRENT BY CONNECTING MULTIMETER BETWEEN NICADS AND D1. INSERT SERIES RESISTOR OR REMOVE SOLAR CELL TO REDUCE CURRENT. D1 KEEPS NICADS FROM DISCHARGING THROUGH SOLAR CELLS (WHEN DARK). SOLAR CELLS ARE FRAGILE. SOLDER AND MOUNT WITH CARE.

☐ LIGHT ACTUATED LATCHING CIRCUITS

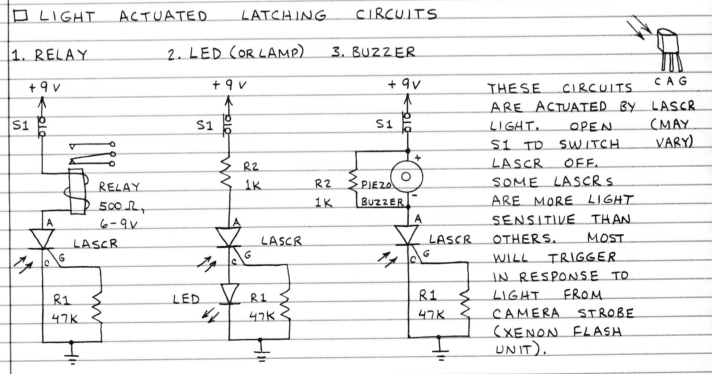

THESE CIRCUITS ARE ACTUATED BY LASCR LIGHT. OPEN S1 TO SWITCH LASCR OFF. SOME LASCRs ARE MORE LIGHT SENSITIVE THAN OTHERS. MOST WILL TRIGGER IN RESPONSE TO LIGHT FROM CAMERA STROBE (XENON FLASH UNIT). (MAY VARY)

DIGITAL IC CIRCUITS

DIGITAL ICs ARE VERY EASY TO USE. HERE'S A SELECTION OF TTL AND CMOS CIRCUITS:

TTL CIRCUITS

☐ OPERATING REQUIREMENTS

1. POWER SUPPLY MUST NOT EXCEED 5.25 VOLTS. SEE PAGE 125 OR USE:

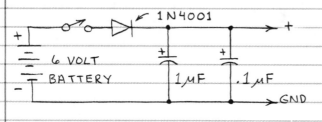

2. INPUTS MUST NOT EXCEED +5.25 V.
3. INPUTS SHOULD ALWAYS GO SOMEWHERE (NOT LEFT "FLOATING").
4. FORCE OUTPUTS OF UNUSED GATES "H" TO SAVE POWER. (EXAMPLE: UNUSED "NAND" — MAKE ONE INPUT H.)
5. AVOID LONG WIRES IN CIRCUITS.
6. CONNECT 1 TO 10 μF CAPACITOR ACROSS POWER LEADS WHERE THEY ENTER CIRCUIT.
7. CONNECT 0.1 μF CAPACITOR ACROSS POWER PINS OF EACH TTL CHIP IN MULTI-CHIP CIRCUITS.
8. REMEMBER, TTL USES MUCH MORE CURRENT THAN LS OR CMOS.

☐ "D" FLIP-FLOP

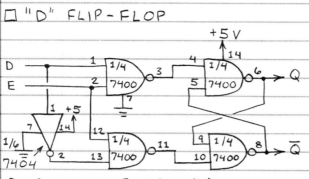

Q = D WHEN ENABLE (E) IS HIGH. NO CHANGE WHEN E IS LOW.

☐ CLOCKED "RS" FLIP-FLOP

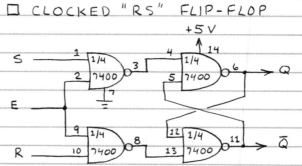

FUNCTIONS AS "RS" FLIP-FLOP WHEN ENABLE (E) IS HIGH.

☐ DUAL LED FLASHER

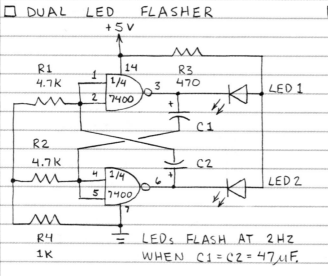

LEDs FLASH AT 2 HZ WHEN C1 = C2 = 47 μF.

☐ TONE GENERATOR

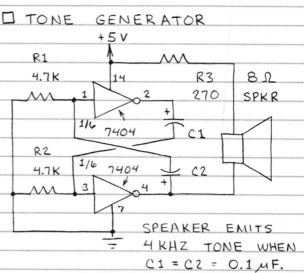

SPEAKER EMITS 4 KHZ TONE WHEN C1 = C2 = 0.1 μF.

☐ 0 TO 9 SECOND (OR MINUTE) TIMER

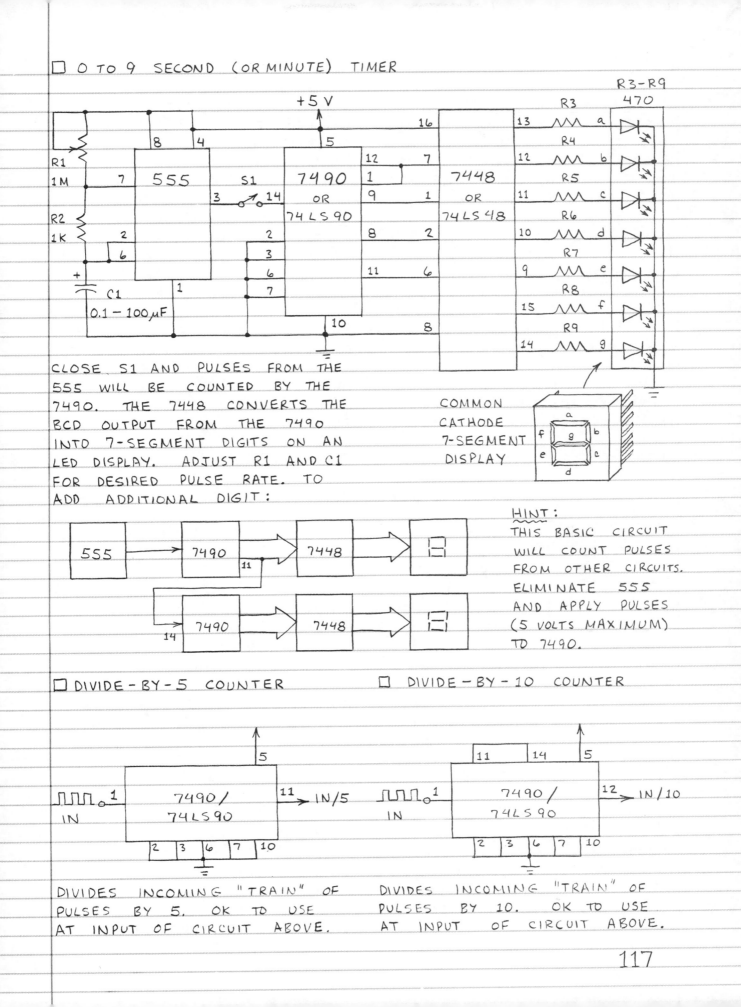

CLOSE S1 AND PULSES FROM THE 555 WILL BE COUNTED BY THE 7490. THE 7448 CONVERTS THE BCD OUTPUT FROM THE 7490 INTO 7-SEGMENT DIGITS ON AN LED DISPLAY. ADJUST R1 AND C1 FOR DESIRED PULSE RATE. TO ADD ADDITIONAL DIGIT:

COMMON CATHODE 7-SEGMENT DISPLAY

HINT: THIS BASIC CIRCUIT WILL COUNT PULSES FROM OTHER CIRCUITS. ELIMINATE 555 AND APPLY PULSES (5 VOLTS MAXIMUM) TO 7490.

☐ DIVIDE-BY-5 COUNTER   ☐ DIVIDE-BY-10 COUNTER

DIVIDES INCOMING "TRAIN" OF PULSES BY 5. OK TO USE AT INPUT OF CIRCUIT ABOVE.

DIVIDES INCOMING "TRAIN" OF PULSES BY 10. OK TO USE AT INPUT OF CIRCUIT ABOVE.

CMOS CIRCUITS

☐ OPERATING REQUIREMENTS

1. THE POSITIVE POWER TO A CMOS CHIP (V_{DD}) CAN RANGE FROM +3 TO +15 (OR +18) VOLTS. USE POWER SUPPLIES ON PAGE 125 OR A BATTERY (BEST):

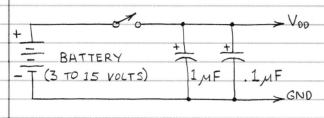

2. INPUTS MUST <u>NOT</u> EXCEED V_{DD}.
3. <u>ALL</u> UNUSED INPUTS MUST GO TO V_{DD} OR GND (⏚).
4. NEVER APPLY AN INPUT SIGNAL TO AN UNPOWERED CMOS CIRCUIT.

☐ HANDLING PRECAUTIONS

1. PLACE CMOS ICs PINS DOWN ON AN ALUMINUM FOIL SHEET OR TRAY WHEN THEY ARE NOT IN A CIRCUIT OR PROPERLY STORED.
2. NEVER STORE CMOS ICs IN <u>NON</u>CONDUCTIVE PLASTIC "SNOW," TRAYS, BAGS OR FOAM. PLUG THEM IN CONDUCTIVE FOAM OR FOAMED PLASTIC WRAPPED IN ALUMINUM FOIL.
3. AVOID USING AN AC LINE POWERED IRON TO SOLDER PINS OF CMOS ICs. USE IC SOCKETS, WIRE-WRAP OR BATTERY POWERED IRON.
4. DRAIN STATIC CHARGE ON YOUR BODY BY TOUCHING GROUNDED OBJECT.

☐ BOUNCELESS SWITCH

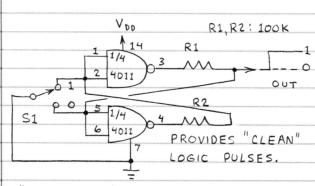

PROVIDES "CLEAN" LOGIC PULSES.

☐ SINGLE LED "GATED" FLASHER

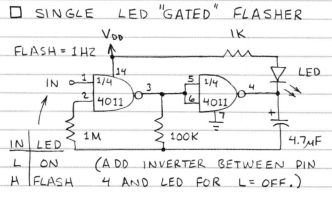

IN	LED
L	ON
H	FLASH

(ADD INVERTER BETWEEN PIN 4 AND LED FOR L = OFF.)

☐ "ONE-SHOT" TOUCH SWITCH

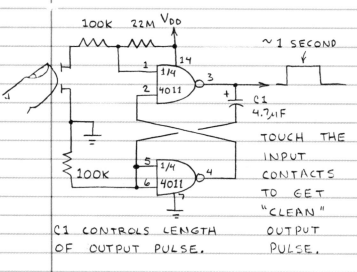

TOUCH THE INPUT CONTACTS TO GET "CLEAN" OUTPUT PULSE.

C1 CONTROLS LENGTH OF OUTPUT PULSE.

☐ DUAL LED FLASHER

LEDs FLASH ALTERNATELY AT ~1 HZ.

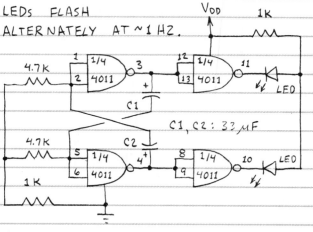

C1, C2: 33 μF

CHANGE C1 AND C2 TO CHANGE RATE.

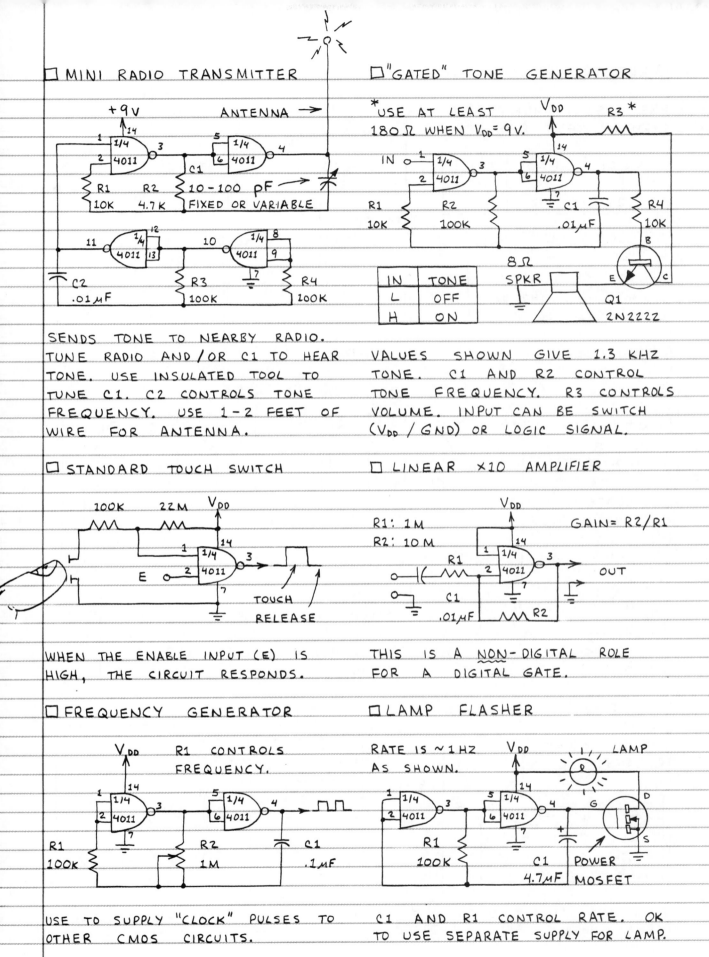

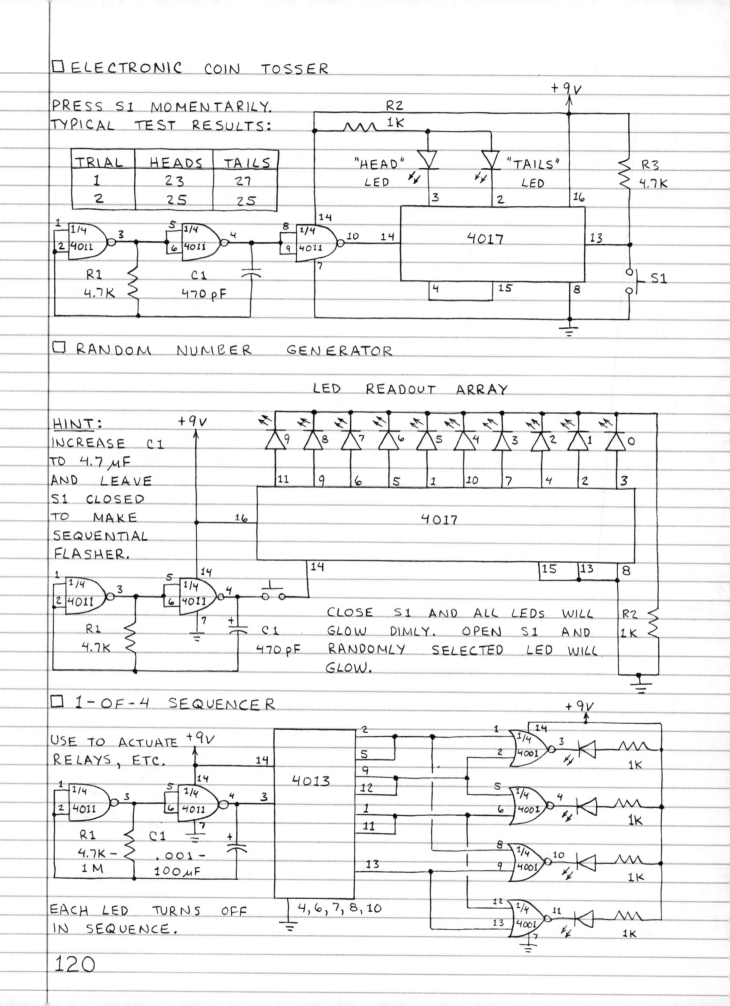

☐ MODEL RAILROAD CROSSING FLASHER LIGHTS

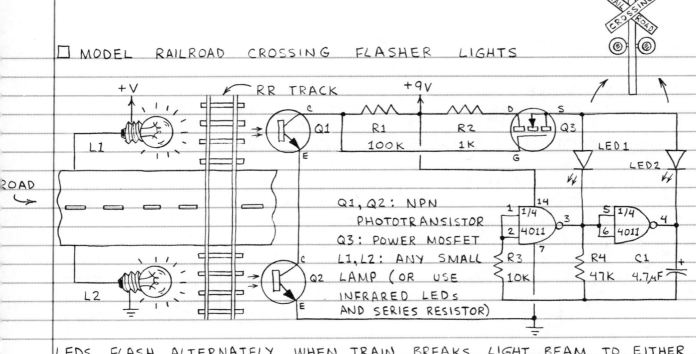

Q1, Q2: NPN PHOTOTRANSISTOR
Q3: POWER MOSFET
L1, L2: ANY SMALL LAMP (OR USE INFRARED LEDs AND SERIES RESISTOR)

LEDS FLASH ALTERNATELY WHEN TRAIN BREAKS LIGHT BEAM TO EITHER Q1 OR Q2. LEDS CONTINUE FLASHING UNTIL TRAIN PASSES. SHIELD Q1 AND Q2 FROM ROOM LIGHTS WITH 1" HEAT SHRINK TUBING.

☐ PROGRAMMABLE GAIN OPERATIONAL AMPLIFIER

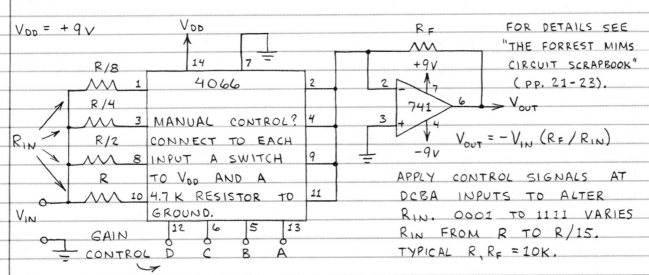

FOR DETAILS SEE "THE FORREST MIMS CIRCUIT SCRAPBOOK" (pp. 21-23).

$V_{OUT} = -V_{IN}(R_F/R_{IN})$

APPLY CONTROL SIGNALS AT DCBA INPUTS TO ALTER R_{IN}. 0001 TO 1111 VARIES R_{IN} FROM R TO R/15. TYPICAL R, R_F = 10K.

☐ ALL ON — ALL OFF SEQUENCER

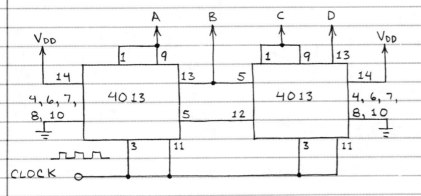

ALL OUTPUTS GO LOW, THEN HIGH, IN SEQUENCE (A...B...C...D...A...B... ETC.). USE WITH LEDs FOR EYE-CATCHING DISPLAY. FOR "BUCKET-BRIGADE" OPERATION, CONNECT PIN 5 OF FIRST 4013 TO PIN 13 (NOT 12) OF SECOND 4013.

LINEAR IC CIRCUITS

YOU CAN MAKE AN AMAZING VARIETY OF CIRCUITS WITH LINEAR ICs. HERE ARE A FEW OF THE MANY POSSIBILITIES:

OPERATIONAL AMPLIFIER (OP-AMP) CIRCUITS

☐ AUDIO AMPLIFIER

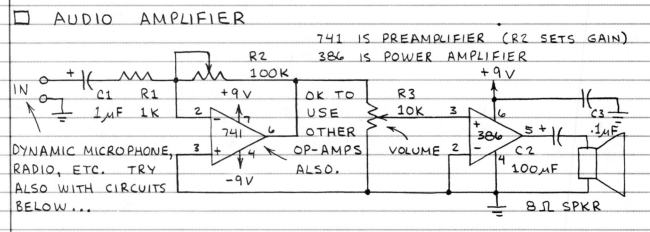

741 IS PREAMPLIFIER (R2 SETS GAIN)
386 IS POWER AMPLIFIER

DYNAMIC MICROPHONE, RADIO, ETC. TRY ALSO WITH CIRCUITS BELOW...

OK TO USE OTHER OP-AMPS ALSO.

☐ MIXER

☐ DIFFERENCE AMPLIFIER

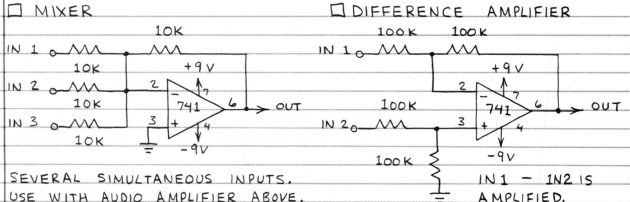

SEVERAL SIMULTANEOUS INPUTS. USE WITH AUDIO AMPLIFIER ABOVE.

IN1 − IN2 IS AMPLIFIED.

☐ LIGHTWAVE VOICE TRANSMITTER*

MIC IS CRYSTAL OR ELECTRET MICROPHONE. LED IS INFRARED TYPE. USE LENS TO FOCUS LIGHT FROM LED INTO NARROW BEAM. TO TEST, PLACE RADIO EARPHONE NEAR MICROPHONE. ADJUST R1 AND R6 FOR BEST RECEIVER SOUND.

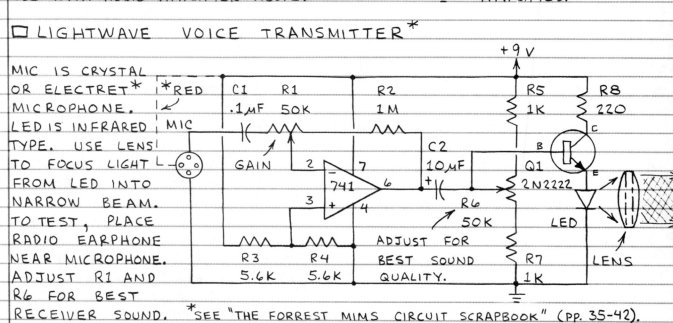

*RED

ADJUST FOR BEST SOUND QUALITY.

*SEE "THE FORREST MIMS CIRCUIT SCRAPBOOK" (pp. 35-42).

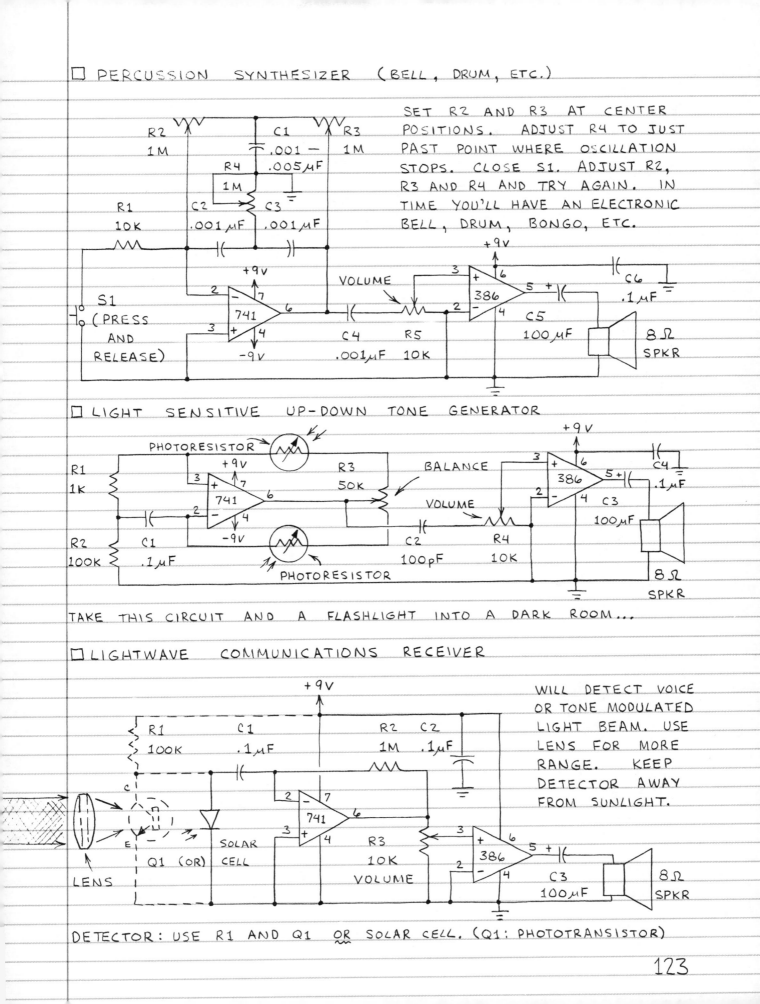

COMPARATOR CIRCUITS

☐ VOLTAGE MONITOR

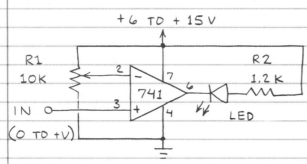

WHEN THE INPUT VOLTAGE IS 0, THE LED GLOWS. THE LED SWITCHES OFF WHEN THE INPUT VOLTAGE RISES TO A LEVEL DETERMINED BY R1. EXCHANGE CONNECTIONS TO PINS 2 AND 3 TO REVERSE OPERATING MODE.

☐ LIGHT LEVEL INDICATOR

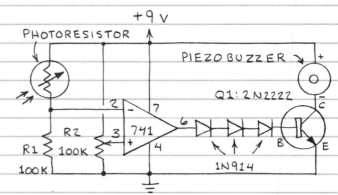

BUZZER SOUNDS WHEN LIGHT FALLS TO LEVEL DETERMINED BY R2. REVERSE CONNECTIONS TO PINS 2 AND 3 TO SOUND TONE WHEN LIGHT INCREASES.

☐ BARGRAPH VOLTAGE INDICATOR

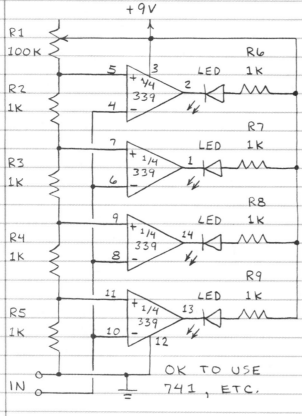

OK TO USE 741, ETC.

LEDs GLOW IN SEQUENCE AS INPUT VOLTAGE RISES. R1 CONTROLS SENSITIVITY.

☐ "WINDOW" COMPARATOR

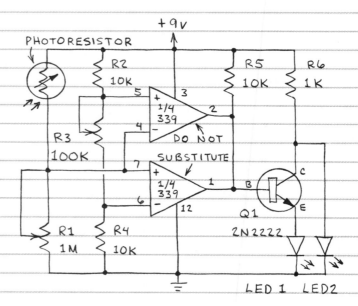

SET R1 TO CENTER POSITION. TURN OFF LIGHTS AND ROTATE R3 JUST PAST POINT WHERE LED 2 GLOWS. CIRCUIT WILL THEN RESPOND LIKE THIS: (R1 & R3 CONTROL RESPONSE)

	DARK	→	LIGHT
LED 1 =	OFF	←– ON –→	OFF
LED 2 =	ON	←– OFF –→	ON

VOLTAGE REGULATOR CIRCUITS

☐ FIXED OUTPUT LINE POWERED SUPPLY

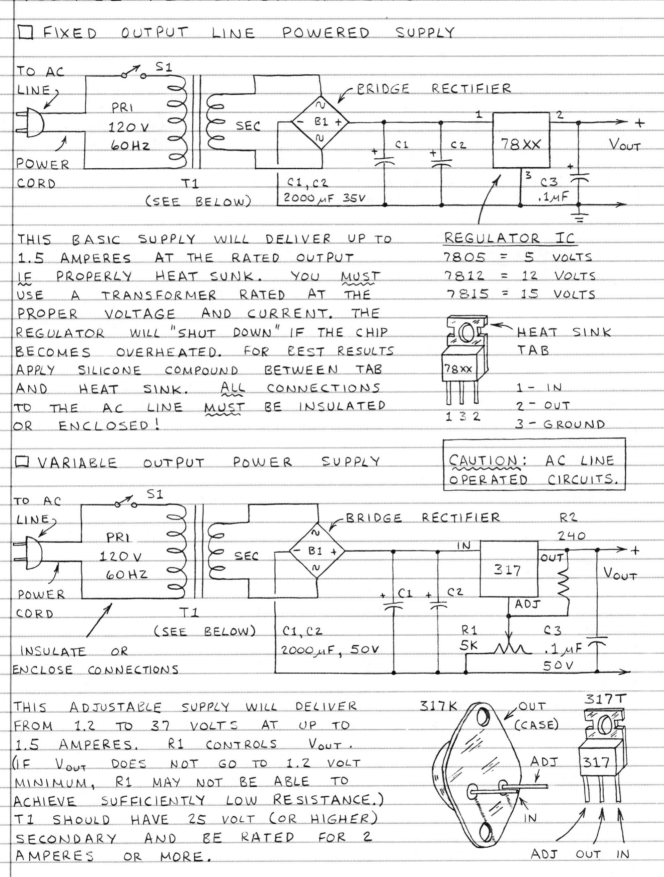

This basic supply will deliver up to 1.5 amperes at the rated output if properly heat sunk. You MUST use a transformer rated at the proper voltage and current. The regulator will "shut down" if the chip becomes overheated. For best results apply silicone compound between tab and heat sink. ALL connections to the AC line MUST be insulated or enclosed!

REGULATOR IC
7805 = 5 VOLTS
7812 = 12 VOLTS
7815 = 15 VOLTS

HEAT SINK TAB
1 – IN
2 – OUT
3 – GROUND

CAUTION: AC LINE OPERATED CIRCUITS.

☐ VARIABLE OUTPUT POWER SUPPLY

INSULATE OR ENCLOSE CONNECTIONS

This adjustable supply will deliver from 1.2 to 37 volts at up to 1.5 amperes. R1 controls V_{out}. (If V_{out} does not go to 1.2 volt minimum, R1 may not be able to achieve sufficiently low resistance.) T1 should have 25 volt (or higher) secondary and be rated for 2 amperes or more.

317K OUT (CASE) ADJ IN

317T ADJ OUT IN

125

TIMER CIRCUITS

☐ BASIC TIMER

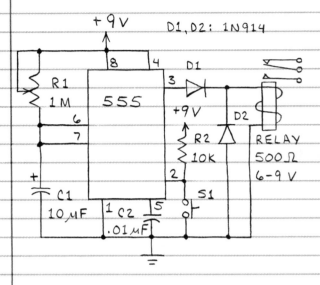

☐ TONE BURST GENERATOR

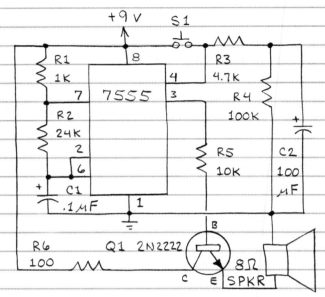

PRESS S1 MOMENTARILY TO START TIMING CYCLE. RELAY WILL BE ACTUATED (PULLED IN) UNTIL CYCLE IS COMPLETE. R1 AND C1 CONTROL LENGTH OF TIME DELAY. USE VERY LARGE VALUE FOR C1 TO GET LONG DELAYS. CIRCUIT WILL RESPOND TO LOGIC PULSES, TOO.

PRESS S1 AND THE SPEAKER EMITS A TONE. RELEASE S1 AND THE TONE CONTINUES FOR SEVERAL SECONDS. C2 AND R4 CONTROL DELAY. C1 CONTROLS FREQUENCY. (USE 7555 ONLY. 555 USES TOO MUCH CURRENT.)

☐ PULSE GENERATOR

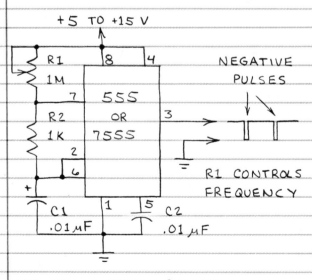

☐ LED TONE TRANSMITTER

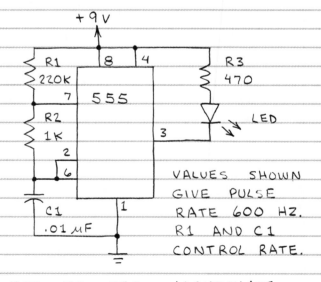

USE TO SUPPLY PULSES TO DIGITAL LOGIC CIRCUITS, ETC.

USE TO TEST LIGHTWAVE RECEIVERS.

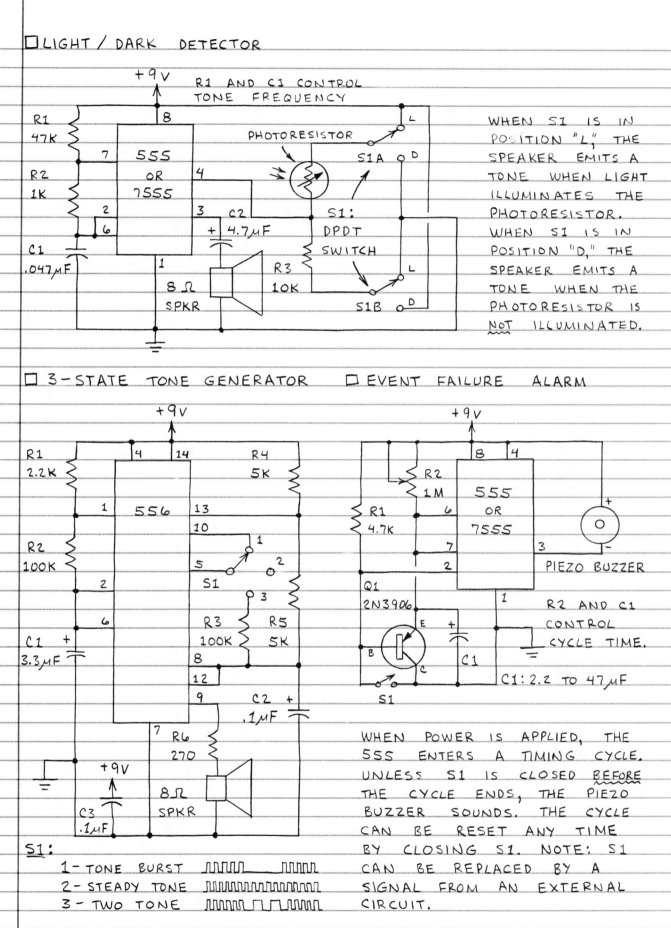

INDEX

A ALTERNATING CURRENT, 18, 36
AMPLIFIER, 92-93, 95, 107, 119, 122
ATOM, 8
B BATTERY, 17, 44
BINARY, 81, 89
BIPOLAR, 48, 51, 104-105
C CAPACITOR, 32-35, 36-37
CIRCUIT, 20-21, 82, 90
CMOS, 91, 118-121
COIL, 38-39
COMBINATIONAL, 86-87, 90
COMPARATOR, 93, 124
CONDUCTOR, 12, 43
CURRENT, 9, 13, 14, 45, 59, 67
D DETECTOR, 70-77
DIFFERENTIATOR, 37
DIGITAL, 79, 80, 83, 91, 116-121
DIODE, 44-47, 78, 82, 101-103
DIRECT CURRENT, 14, 36
E ELECTROMAGNET, 15
ELECTROMAGNETIC SPECTRUM, 63
ELECTRON, 8-9, 42-43, 62, 66, 70
ELECTROSCOPE, 11-12
F FIELD-EFFECT, 52-56, 92
FLIP-FLOP, 88-89, 94, 116
G GATE, 80, 82-87
GROUND, 21
I INSULATOR, 12, 43
INTEGRATED CIRCUIT, 78-95
INTEGRATOR, 37
INVERSE SQUARE LAW, 65
ION, 8
J JUNCTION FET, 52-53, 56, 106
L LASCR, 76
LENS, 65, 122, 123
LIGHT, 62, 63, 66-68, 77, 111, 114, 122
LIGHT EMITTING DIODE, 46, 66-69, 112
LINEAR (IC), 79, 92, 95
M MAGNETIC FIELD, 13
METER, 13, 25, 114
MICROPHONE, 27
MOSFET, 54-55, 56, 107
MULTIMETER, 19
N NOISE, 22-23
NUCLEUS, 8

O OHM'S LAW, 14, 103
OPERATIONAL AMPLIFIER, 93, 122-123
OPTICAL COMPONENTS, 64
OPTICAL SPECTRUM, 63
P PARALLEL, 20-21, 31, 35, 69, 77
PHOTODIODE, 72-73
PHOTON, 62, 66, 70, 72, 74, 76
PHOTORESISTOR, 70-71, 114-115, 124
PHOTOTHYRISTOR, 76, 115
PHOTOTRANSISTOR, 74-75, 114, 121, 123
PN JUNCTION, 44, 66, 70, 72
POWER, 14, 116, 125
PROTON, 8
PULSE, 22-23, 119, 126
R RC CIRCUITS, 37, 94, 100
RELAY, 25, 75, 114, 115
RESISTANCE, 14
RESISTOR, 28-31, 69, 78
S SAFETY, 19, 35, 98, 100, 111, 125
SCR, 58-59, 110
SERIES, 20-21, 31, 35, 69, 77
SEMICONDUCTOR, 42-43, 62-63, 66, 70
SEQUENTIAL, 88-90
SIGNAL, 22-23
SILICON, 42-43, 72
SINE WAVE, 18
SOLAR CELL, 77-115
SOLDER, 98
SOLENOID, 15
SPEAKER, 27
STATIC ELECTRICITY, 10, 106, 118
SWITCH(ES), 25, 26, 58, 80
T THERMOELECTRIC, 17
THYRISTOR, 58-61, 110-111
TIMER, 93, 106-107, 117, 126
TRANSFORMER, 40-41
TRANSISTOR, 48-57, 74-75, 78, 83, 104
TRIAC, 60-61, 111
TTL, 91, 116-117
U UNIJUNCTION, 57, 108-109
V VOLTAGE, 14, 45, 68
VOLTAGE REGULATOR, 95, 103, 125
W WAVE(S), 22-23
WIRE, 24
WIRE-WRAP, 97